MAPPA MUNDI
THE HEREFORD WORLD MAP

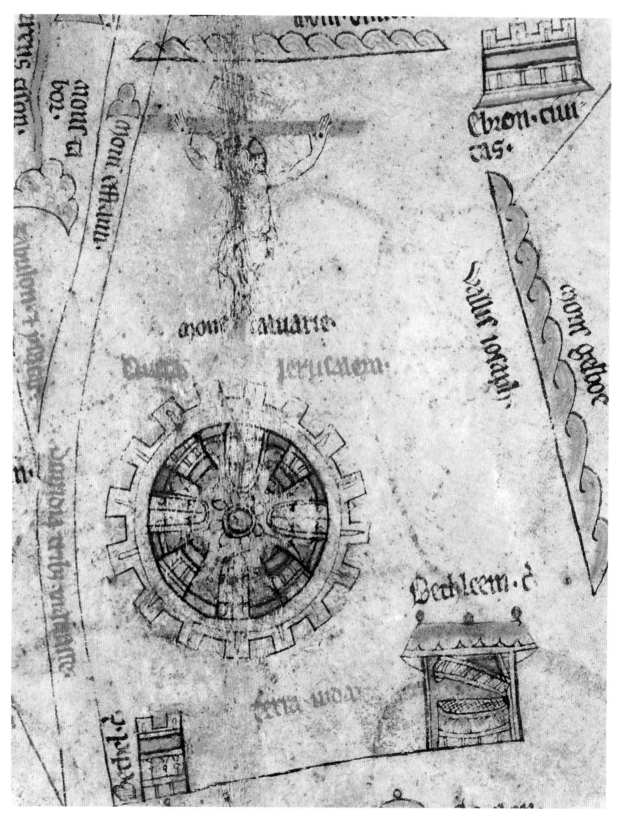

JERUSALEM ON THE HEREFORD MAP

Above the city is a drawing of the Crucifixion. Jerusalem is the exact centre of the map, and the compass point in the middle may have been used to draw its circular outer frame as well as the circle of the city walls. This, however, was no more than cartographic convention – it does not mean that medieval geographers saw Jerusalem as the centre of the world.

MAPPA MUNDI
THE HEREFORD WORLD MAP

P.D.A. HARVEY

UNIVERSITY OF TORONTO PRESS

TORONTO AND BUFFALO

PREFACE

The Select Bibliography is intended to take the place of many footnote references; when reference is made to a book listed there it is identified by the author's name alone.

It is a privilege that I much appreciate to have been invited by the Hereford Mappa Mundi Trustees to write this book and to have been given the opportunity of examining the map outside its case. In the course of the work many people have given me substantial help by way of advice and discussion, answering enquiries and generous hospitality, and I am especially grateful to Mr P. Barber, Dr A.D. Baynes-Cope, Mrs A. Butterworth, Canon J. Butterworth, Mr F. Herbert, Mr D. Harbour, Miss D.S. Hubbard, Dr M. Kupfer, Mr I. Milford, Mrs B. Morgan, Mr D.E. Morgan, Mrs A. Payne, Mme M. Pelletier, Dr N. Ramsay, Mr N. Skyrme, Dr S.D. Westrem and Miss J. Williams. I would add a special word of thanks to those with whom I have particularly worked, at Hereford Canon John Tiller and at the British Library Ms Kathleen Houghton and Mr David Way.

P.D.A.H.

© 1996 P.D.A. Harvey
First published 1996 by
The British Library
Great Russell Street, London WC1B 3DG

Published in North America by
University of Toronto Press Incorporated

Canadian Cataloguing in Publication Data
is available from the publisher

ISBN 0-8020-0985-9 (cloth)
ISBN 0-8020-7945-8 (paper)

Designed by John Mitchell
Set in Caslon by
Nene Phototypesetters, Northampton
Colour origination by York House Graphics, London
Printed in Great Britain by
The Clifford Press, Coventry

CONTENTS

1 · The map and its history

THE MAP DESCRIBED

Hereford Cathedral's medieval map of the world is one of the most interesting artefacts to survive from thirteenth-century England, but at first sight it is far from prepossessing. It has aged over the centuries, and its pattern mostly of various shades of brown and black has a dull, uninteresting appearance. But when it was made, seven hundred years ago, it was alive with colour. The parchment on which it is drawn will have been whiter then, and the dark brown seas that dominate the whole map were a bright green. It is easier to see that the rivers were once blue, probably bright blue, for though much of the colour has flaked off, leaving only the brown ground that held it to the parchment, it is in places almost intact and some traces remain throughout. The gold leaf has fared better: much of it remains, but it was austerely confined to some of the lettering – the names of the three continents and India on the map, the names of the four cardinal points in a circle

Left:

THE HEREFORD WORLD MAP
The map is drawn on a single piece of parchment, 5 feet 2 inches high and 4 feet 4 inches wide (1.58 by 1.33 metres). Its colours have changed over seven hundred years – the dark brown seas were originally bright green, the rivers blue. The large gold letters are in a different hand from the rest, and this may explain why the names of two continents have been reversed – 'AFFRICA' appears on the left, 'EUROPA' on the right. In the top right both the Red Sea and the Persian Gulf are coloured red.

Right:

LAYOUT OF THE HEREFORD MAP
Drawn as a circle with east at the top, Jerusalem in the centre, this was a conventionalised form of the way the thirteenth century envisaged the inhabited world. The earth was in fact seen as a sphere, of which the known, inhabited part occupied much of the northern half – no part of this inhabited world extended as far as the equator.

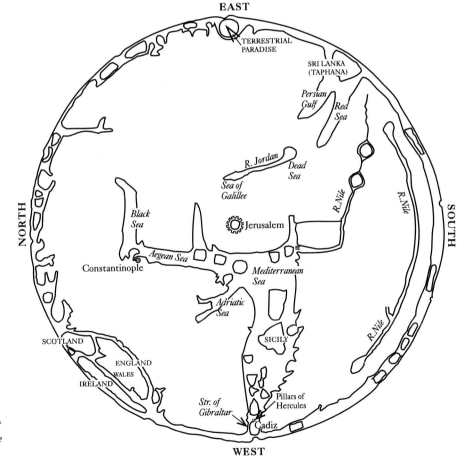

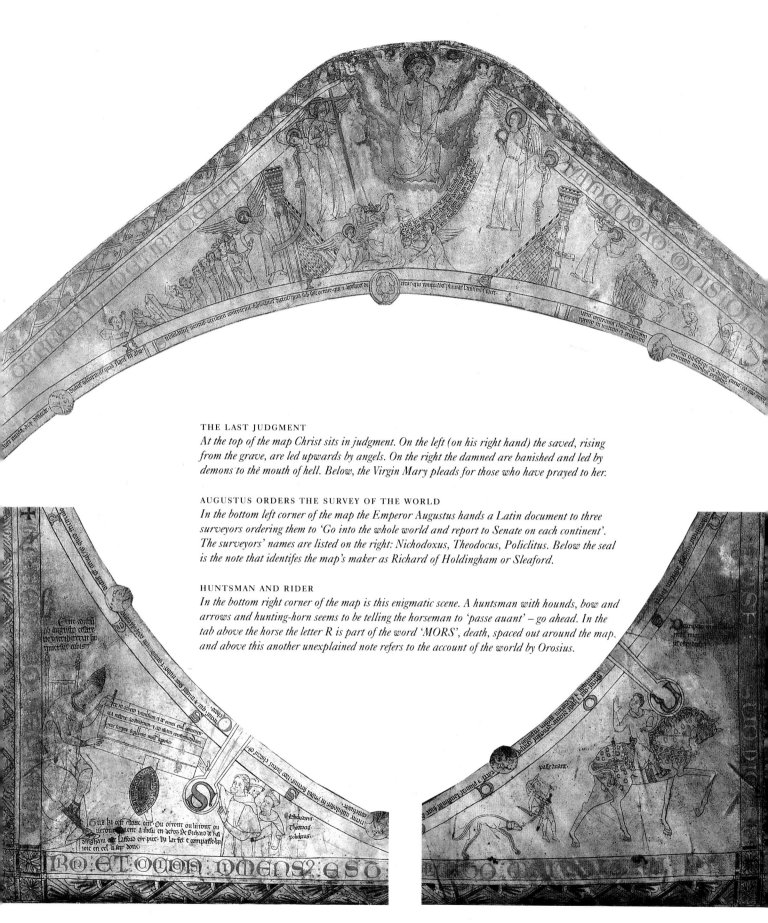

THE LAST JUDGMENT

At the top of the map Christ sits in judgment. On the left (on his right hand) the saved, rising from the grave, are led upwards by angels. On the right the damned are banished and led by demons to the mouth of hell. Below, the Virgin Mary pleads for those who have prayed to her.

AUGUSTUS ORDERS THE SURVEY OF THE WORLD

In the bottom left corner of the map the Emperor Augustus hands a Latin document to three surveyors ordering them to 'Go into the whole world and report to Senate on each continent'. The surveyors' names are listed on the right: Nichodoxus, Theodocus, Policlitus. Below the seal is the note that identifes the map's maker as Richard of Holdingham or Sleaford.

HUNTSMAN AND RIDER

In the bottom right corner of the map is this enigmatic scene. A huntsman with hounds, bow and arrows and hunting-horn seems to be telling the horseman to 'passe auant' – go ahead. In the tab above the horse the letter R is part of the word 'MORS', death, spaced out around the map, and above this another unexplained note refers to the account of the world by Orosius.

around it and the letters MORS in roundels outside. The red no longer stands out as it must once have done, but it is still intact and bright; it is used for some decoration and lettering – including the names of peoples and provinces on the map and the wording round the edge of the parchment – and, as on some other medieval maps, not only for the Red Sea but for the Persian Gulf as well. The ink used for the outlines and most of the wording is mostly still crisp and black, unlike the ink of many contemporary manuscripts. Probably several other colours were used for the shading in the pictures on the map itself and in the borders: what now appear as grey, yellow and olive-brown can be distinguished.[1]

Even now the map is impressive in its size and complexity. It is drawn on a single, extraordinarily large piece of parchment, shaped like the gable end of a house, overall some 5 feet 2 inches high and 4 feet 4 inches wide: 1.58 metres from the apex to the bottom, 1.09 metres from the base of the triangular top to the bottom and from 1.30 to 1.33 metres across. Most of this space is taken up by the map itself in a vast circle; around it are the outer band naming the cardinal points and an inner band identifying the twelve winds of classical authority, the four major winds each shown by a tiny naked human figure with grotesque head, kneeling on all fours and puffing hard, the others by a dragon's head, blowing with open mouth. Above the map is Christ in judgment, with the saved being led to paradise on his right hand and the doomed being led to hell on his left; below him the Virgin Mary, with bared breasts, pleads in four lines of French verse for those who sought her intercession – this and other inscriptions outside the map are given, with translation, in Appendix 1. The world was God's creation, a point brought home by the letters MORS, death, spaced out beyond its bounds. Below the map on the left the Emperor Augustus gives a sealed document of wholly thirteenth-century appearance to three named surveyors, ordering them to measure the world, and an inscription in red round the edge of the whole parchment tells how one of them, Nichodoxus, measured the east, Policlitus the south, and Theodocus the north and west – inconsistently, it says this was by order of Julius Caesar. Below the map on the right a man on horseback, hand raised in salutation, passes a huntsman with bow and arrow, horn and a pair of hounds, who is apparently saying 'passe avant' (go ahead)[2] – perhaps one of the Roman surveyors (he wears similar clothes) being given priority on the road, perhaps the medieval author of the map in person. It may, however, be a figure typifying the individual living in the confines of the world to which he is gesturing – or, indeed, there may be a topical or personal allusion now lost.

The map itself looks strange to our eyes, though the places that western Europe knew through direct experience are in more or less correct relationship, while often much out of proportion and oddly skewed out of shape. Sea surrounds the whole land-mass. East is at the top, and there, a walled island off the coast of Asia, is the earthly paradise with Adam and Eve taking the forbidden fruit; their expulsion is pictured on the adjacent mainland. At the foot of the map are the Straits of Gibraltar, with the Pillars of Hercules set on the island of Cadiz. Above is the Mediterranean, studded with islands – the triangular Sicily among them – and stretching nearly to the centre of the map; at the sea's eastern end the Nile flows in from the south, the Adriatic lies to the

TYPICAL DETAILS
In the parts of the map covering – roughly – the ancient Roman empire little is shown beyond rivers, mountains, provincial boundaries and towns. Elsewhere is an encyclopaedic mass of information about the people, the history and the natural history of distant lands – all pictured as well as described.

CONSTANTINOPLE AND THE BLACK SEA
The walls and three towers of Constantinople are pictured facing the sea, and thus upside down. To the left is the Danube delta, and beyond it the River Dnieper (fl' Danaper).

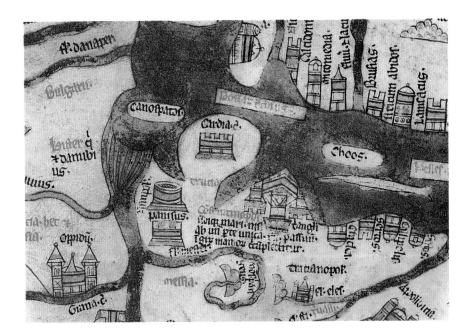

SAMARCAND AND THE RIVER OXUS
Samarcand is the city at the top, the Oxus the river on the left and the Bactrus the river at the bottom. Here the pelican revives its young with the blood from its own breast, and on the left are the bird-like people called the cicone.

4

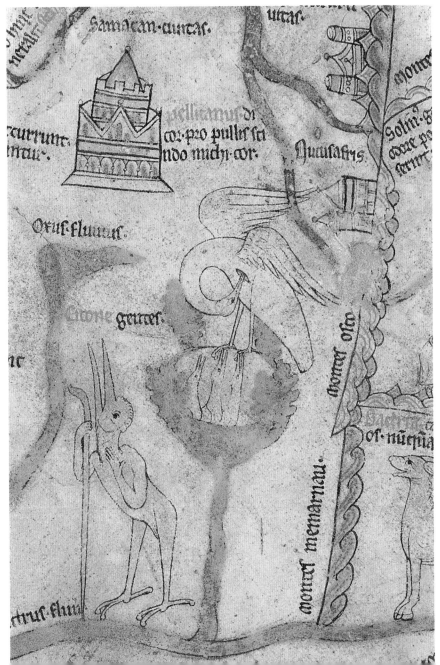

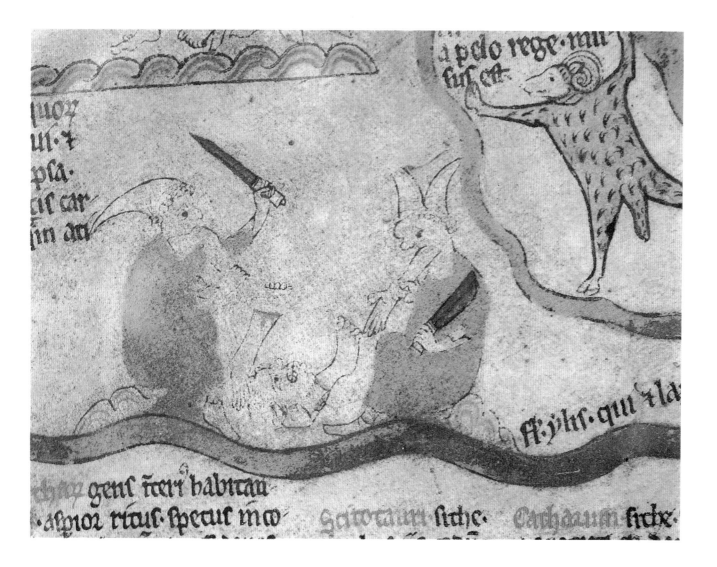

PART OF NORTHERN ASIA
The essedones *are a people who eat the corpses of their parents in solemn feasting, thinking this better than letting them be consumed by worms. To the right is the golden fleece, from the ancient Greek legend of the Argonauts. On the rivers are traces of the blue colouring that has mostly flaked away.*

north-west and a long channel leads northwards. This is the Aegean; beyond the straits at Constantinople it makes a right-angled turn eastwards as the Black Sea. At the centre of the map is Jerusalem, a circular walled city with the Crucifixion above it, and, further east, the Jordan with the Dead Sea and Galilee. To the south-east are the Red Sea and Persian Gulf and, disconcertingly, the island of Sri Lanka (*Taphana*) lies at their mouth. There are in fact islands all the way round the land-mass, and by far the largest is England and Wales in the north-west – Scotland as well as Ireland is a separate island. Parallel to the outer coastline at the south of the map, a long river with a lake at each end effectively separates southern Africa from the rest of the continent. This is the upper Nile; though it does not visibly connect with the river that flows into the Mediterranean, it was taken to be the same river, flowing underground for part of its length. Rivers, though denser in Europe than elsewhere, are shown throughout the map, and so are mountain chains, isolated peaks and boundary lines between peoples and provinces.

All the land on the map is covered with writing and pictures.

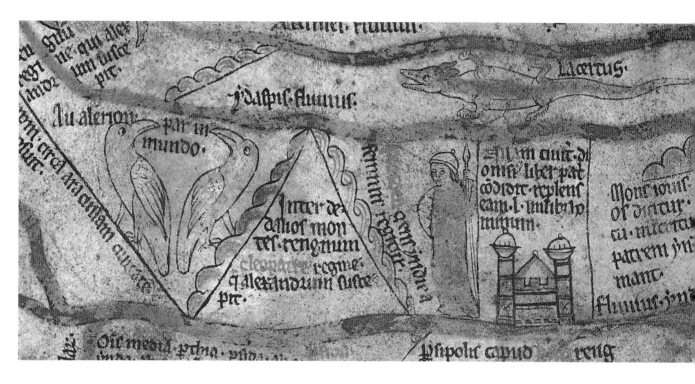

Above:

PART OF INDIA

On the left are the birds called avalerion. *There is only a single pair, which after sixty years produces two eggs; when they hatch the parents drown themselves. On the right is a woman warrior from a people governed by women.*

Right:

LINCOLN AND ITS SURROUNDS

Lincoln is shown in more realistic detail than any other town on the map, suggesting strongly that it was made in that area. From the cathedral or castle on top of the hill a street lined with houses runs down to the River Witham. Below are Northampton and Ely.

Left:

HEREFORD AND ITS SURROUNDS

Hereford was not on the map when it was originally drawn, but it was added not long after by a different hand, along with the River Wye. These changes necessitated rewriting the name of the River Severn, originally probably fl' Sabrina *but now* Sabrina fl'. *Probably at the same time the hill from which both Severn and Dee took their source was named Clee Hill,* Mons Clece.

However, west and south-east Europe, Asia Minor, Palestine and the north African coast differ from the rest – as we shall see, it may be significant that this is broadly the area of the old Roman Empire. Here it is mostly just the cities that are shown, conventionally pictured and named – indeed, they are so numerous that there is hardly room for anything else. Elsewhere there are a few towns, but otherwise the space is taken up with innumerable snippets of information, most of them accompanied by a picture. Some are simply geographic – the note on Cyprus gives its dimensions, 180 miles long and 120 miles wide. Some are biblical, historical or literary, recalling events that occurred in the past. But more tell us what we would find in these distant places – strange animals, birds and plants, people with weird customs, monstrous races of peculiar physique. The whole forms a kind of encyclopaedia, arranged geographically. On the map itself all the wording is in Latin – it is only a few of the inscriptions outside the border that are in French. French was at this time widely spoken by the English upper classes, and its use here suggests that the map's author envisaged a readership beyond the learned and scholarly circles for whom Latin alone was appropriate.

THE MAKING OF THE MAP[3]

Unlike most medieval maps, this one names its author. A note in French in the lower left corner, below the document and seal of Augustus, asks all who hear, read or see the map, which it calls *cest estorie*, this story or picture, to pray for Richard of Holdingham (*Haldingham*) or Sleaford (*Lafford*) who made it.[4] Holdingham is a village just outside Sleaford, in Lincolnshire, and the map itself shows that the author was connected with Lincoln: the city is marked with a picture more elaborate than any other in England, and unlike any other it shows its actual topography with the cathedral or castle on the hill and a house-lined street leading up to it from the River Witham. However, the map shows too that soon after its completion it found, unexpectedly, a home at Hereford. Hereford and the River Wye are clear additions to the map, drawn less neatly than the rest and necessitating re-writing the name of the Severn (*Sabrina fl'*), and probably at the same time the mountain from which both Severn and Dee flow was named Clee Hill (*Mons Clece*).[5] There is also internal evidence of the map's date. Almost certainly Conway and probably also Carnarvon were on the map as originally drawn and neither would have appeared before Edward I began work on his castles there, at Conway in 1277 and at Carnarvon in 1283.

Turning to other sources to expand what we learn from the map itself, we find that one Richard of Battle (*de la Bataylle* in French, *de Bello* in Latin), in 1260 a rector in Kent, was a canon of Lincoln by 1265 and prebendary of Sleaford there from at latest 1277; he had ceased to hold the prebend by 1284. He is not named Holdingham, but if Battle was his family name Holdingham might well have been where he was born or lived, and such an alternative surname would be normal at this time. The only surviving household accounts of a medieval bishop of Hereford, in 1289-90, record gifts of venison to Richard of Battle and of

RULED LINES FOR AN INSCRIPTION
On some of the map's longer inscriptions we can see guide lines ruled above and below each line of writing. Here they appear on the note marking the start of the route of the Exodus at Rameses, on the left. The town encircled by the River Nile is Cairo, called the new Babylon, Babilonia, as usual in the medieval west.

SEQUENCE OF ENTRIES
Sometimes the spacing of entries on the map reveals the order in which they were written. Here the note in red on the boundary between Asia and Africa must have been written before the black note below, identifying Alexander's camp, as its second line would otherwise have been further to the right.

money to his servant, and from 1305 until his death in 1326 Richard of Battle was a non-residentiary canon of Hereford and prebendary of Norton. Meanwhile, besides acquiring several other benefices else-where, Richard of Battle had had a notable career in the diocese of Salisbury, becoming a canon there in 1298, the bishop's vicar-general in 1299 and keeper of the spiritualities when the see was vacant in 1315.

All this fits extraordinarily well with the map's own evidence of its origins. However, in 1957 two scholars, N. Denholm-Young and A.B. Emden, argued, independently, that this was the career not of one man but of two, one a canon of Lincoln and the other a canon of Salisbury and Hereford; the two may of course have been related. In 1974 Dr W.N. Yates, eliminating some of the difficulties raised in 1957, sug-gested that there is no insuperable objection to seeing Richard of Battle as one man, who lived to be at least eighty-six – a long, but not impos-sible lifespan even in the fourteenth century. However, as the anniver-sary of his death – 4 November, year unstated – was recorded and commemorated at Lincoln, it is unlikely that he died more than forty years after leaving, and in fact his executors in Kent are mentioned in 1279.[6] We should thus see the canon of Hereford and Salisbury as a second Richard of Battle. If the two were closely related this might

9

explain how a map produced by the elder Richard at Lincoln passed to Hereford with the younger; but alternatively it could mean that the younger Richard came from the Lincoln area, that it was he who was Richard of Holdingham and that the canon of Lincoln had nothing to do with the map – this would in fact better fit the map's date of 1283 or later.

However well it seems to fit, the identification of the map's Richard of Holdingham with either Richard of Battle is entirely circumstantial. Dr Yates rightly warns us against taking it for granted, and mentions, as another possible candidate, Richard of Sleaford, rector of Normanby-le-Wold, 17 miles from Lincoln, from 1283 to 1294 – though he had no known contact with Hereford. All we can say is that it seems likely, but far from certain, that one Richard of Battle, canon of Lincoln or canon of Hereford and Salisbury, was the same as Richard of Holdingham named on the map.

Nor is it certain what part Richard of Holdingham played in its creation. According to the inscription on the map he *lat fet e compasse*. We might translate this as 'designed and executed it', but this cannot be reconciled with the evidence of the map itself. Apart from the gold lettering, the whole map was originally written and drawn by a single hand.[7] This was the hand of a craftsman and artist, but it was not the hand of a scholar, for some of the few mistakes are not mere slips of the pen – they arise from a failure to understand what was being copied.[8] It follows that the map was designed and executed by different persons. Whether Richard was the craftsman or the author – whether he wrote and drew the map in his own hand from someone else's draft or whether he himself drafted it to have it produced by someone else – we cannot tell.

It is just possible that the map is a faithful copy of an existing finished map which itself contained these few failures to follow its author's meaning, but it is easier to suppose that it was made directly from a draft in which the position of every picture, the length of every inscription, had been worked out with extreme care but with some of the wording left in abbreviated form to be expanded by the copyist.[9] Certainly it was made with great exactitude. There are no signs of second thoughts, of corrections made in the course of writing or of real problems in fitting words or pictures into the available space; a very few inscriptions are slightly constricted – such as one at the east end of Crete – but this is scarcely perceptible and would probably be unavoidable even with the most careful preparation. The detailed drafting was not done on the parchment used for the finished map, though drafting lines are visible on a few of the rivers, as well as ruled lines for the longer inscriptions (above and below the lines of letters) – these all seem to have been drawn in crayon or very pale ink. How the design was transferred to the parchment for the fair copy is not clear; it has not been possible to establish whether the many pin-pricks relate to the design of the map. However, two of the tiny holes in the map were clearly made by compasses to draw circles – one is in the centre of the round labyrinth on Crete and the other at Jerusalem, where it was perhaps used both for the city wall and also, being the centre of the whole map, for the three concentric circles that mark its outer edge.

To judge from the spacing, the outlines and rivers seem to have been drawn first, and the pictures entered before the accompanying wording, and very occasionally we can tell from the spacing that one inscription was entered before another, as on the boundary of Africa and Asia. The letters in gold – the names of the continents and India on the map, the names of the cardinal points around it and MORS outside – differ from the rest and were probably entered last of all by another hand. This helps to explain the glaring error on the map – 'AFFRICA' is written across Europe, 'EUROPA' across Africa – which must have caused the map's author much distress. A very few corrections were made, probably by the original writer after the map's completion. One is in the note on the length and breadth of Africa, both accidentally called *longitudo*, length; the second has had a dot placed below each letter – an accepted neat alternative to crossing out – and 'lat'', *latitudo*, breadth, added.[10] However, the naming of Clee Hill and the changes around Hereford, though contemporary, are probably in a different hand from the map itself. We can only speculate on the significance of this for the identity of Richard of Holdingham and his role in the map's creation.

THE MAP'S SUBSEQUENT HISTORY

It is not only in general ageing and in the loss of some of its colour that the map has suffered over the years. It can be seen that at some point it was folded lengthways. In the lower left corner there is heavy scoring, centred on France but extending into adjacent areas; this has been explained as a late-medieval expression of anti-French feeling,[11] but it could have been accidental – it might have been caused, for instance, by the 'quantity of glass lanthorns' that had been found 'piled against the map' in 1812.[12] In the top part of the map there are four holes in the parchment, one beside the M of MORS, one a little to the south of the island of paradise, one in the Red Sea, and one beside the south-by-south-east wind, from which a crack extends some way along the southern coast of Africa; these holes have all been repaired with parchment patches and the patches on the last two have been painted to avoid disfiguring the map. Just below the last of these holes the inscription beside the horned and beaked satyr has been mostly erased, perhaps deliberately. Finally, right round the map its extreme edge is frayed and has been backed with new parchment.

This damage around the edge of the map can be easily explained. In the nineteenth century it was kept fixed in its case by a thin brass strip along each side, secured by a large number of round-headed brass nails which must have passed through the extreme edge of the parchment.[13]

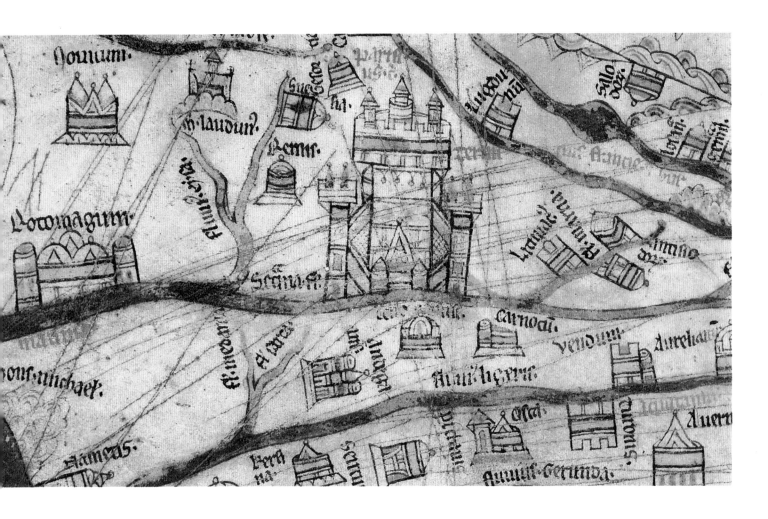

This arrangement, perhaps even these same strips and nails, probably dated from the middle ages, possibly from the time the map first came to Hereford. This would be consistent with the radio-carbon dates of the wood: the boards making up the back panel of the case probably date from between 1270 and 1400 and between 890 and 1020 – perhaps re-using old wood – and the crockets along the top from between 1040 and 1280. It would be consistent too with all we know of the rest of the case, now lost. This case, itself an object of great interest, was recorded in two drawings of the 1770s. It was a triptych, over 3 metres wide with its wings open and the earliest of substantial size known from England. In the centre was the map, and above it was a painted canopy, with carved oak-leaf crockets and a finial; on the centre panel, as a narrow border around the map, scrolled decorations with dragon-like animals were painted in red on a white background. On the wings was painted the Annunciation, the angel on the left wing, the Virgin on the right, both dressed in red with blue undertunics; it is not known what was on the back of the wings, visible when they were closed as doors covering the central panel with the map. From the late eighteenth century onwards the case was gradually dismembered. The two wings were taken off, probably between 1780 and 1800, and seem to have been destroyed or lost. The decorated border to the map, painted on the back panel, was removed at the British Museum – to the dismay of the

Several holes in the map have been repaired with patches of parchment, and two were coloured so as not to disfigure the map. Here, south of the mouth of the Red Sea, the patch has been painted light and dark brown to tone with the map's own parchment and with the ocean. Between the patch and the Red Sea is the burning mountain, mons ardens, *with flames issuing from each side.*

Museum's Keeper of Manuscripts, Sir Frederic Madden – when the map was taken there for restoration in 1855. New doors were added to the case, now glazed and with a curtain in front, in 1868. The finial at the top of the case disappeared between 1931 and 1946, probably when the map was evacuated during the Second World War. Finally, following the map's further restoration – which is probably when the brass strips and nails were lost – a new case was provided in 1948 and the old one thrown out as lumber. In 1989, when its importance was recognised, it was discovered in the cathedral's former stable; it was radiocarbon dated at the expense of *The Observer* newspaper and placed on exhibition along with the map.

What happened to the map when it first came to Hereford, what purpose it was meant to serve, can only be surmised. As we shall see, similar large world maps were painted on the walls of royal residences, probably in the context of other paintings of allegorical and moral import. Of themselves they could be viewed as a form of enlightenment or diversion for the literate, who could read them, or for the illiterate, to whom they could be explained, conveying interesting, even entertaining, information in a learned and pious setting. This is probably how the map was seen at Hereford, and its acquisition may well have been connected with the learning and wide interests of Richard Swinfield, bishop from 1283 to 1317. If the triptych was made for the map very soon after it came to Hereford, the possibility arises that it was used from the start as an altarpiece, a representation of Christ in judgment above the whole of God's earthly creation – in greater detail than any worshipper could see, but this was hardly the point. After the Reformation it will have seemed less appropriate and was removed to the cathedral's library. However, this view of the map's role has been seriously questioned by Dr M. Kupfer, who argues that the early association of map and triptych should not be taken for granted and that in any case the map was quite inappropriate for an altar. Earlier, in the eleventh and early twelfth centuries, some monastic schools used a world map in teaching; it would seem less well suited to the more rigorous, formally structured education of the late thirteenth century, but one possibility is that the map is to be connected with the cathedral school at Hereford, highly developed a century earlier.[14] However, Dr Kupfer suggests that the map might have been used at Hereford for less formal instruction, as an aid not to teaching but to preaching. Its precise role at Hereford – why it was acquired by the cathedral, what purpose it was meant to serve there – can be discovered only from inference and analogy – we have no direct evidence.

The map may well have been in the cathedral library in the early seventeenth century, when the signature of 'John Nicolles' was written beside the horseman in the lower right corner; he may have been the son or other relative of William Nichols, vicar-choral who died in 1635.[15] The ink is in the same colour as the ringed dapplings on the horse, and we might imagine that the young John was allowed to add this decoration and then to sign his work.[16] Certainly the map was in the library at the time of the earliest known reference to it, in a note written by Thomas Dingley of Dilwyn, Herefordshire, in the early 1680s or a little earlier: 'Among other curiosity in this Library are an Map of ye World drawn on Vellum by a Monk Kept in a frame with two

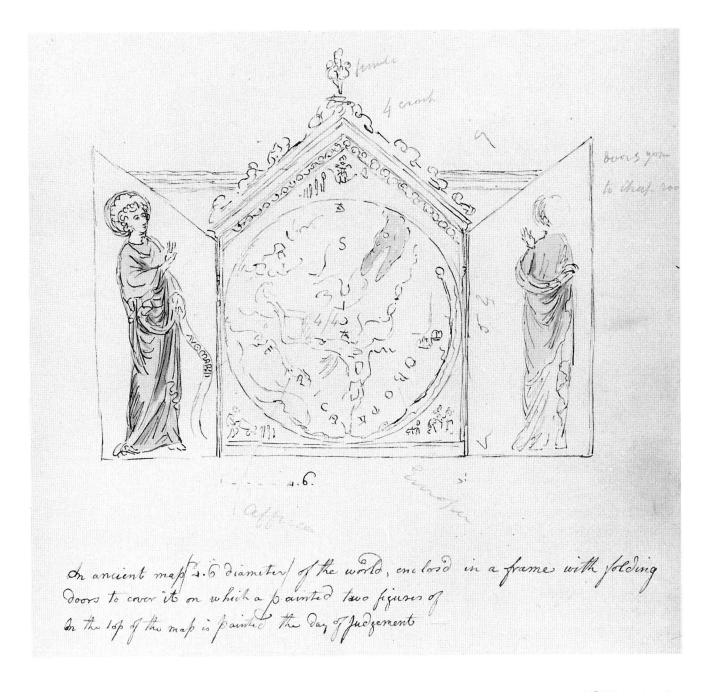

On ancient map 4.6 diameters of the world, enclosd in a frame with folding
doors to cover it on which a painted two figures of
On the top of the map is painted the day of Judgement

THE HEREFORD MAP IN ITS TRIPTYCH,
LATE 18TH CENTURY
*This drawing by John Carter is our best
evidence for the original appearance of the
triptych that then contained the map. The
two painted wings were removed not long
after this drawing was made in the 1770s
and were destroyed or lost.*

British Library, Additional MS. 29942, f.148

doors – with guilded and painted Letters and figures'.[17] This closely
matches the opening words of the earliest reference in print, by Richard
Gough a century later: 'In the library of Hereford cathedral is preserved
a very curious map of the world, inclosed in a case with folding doors,
on which are painted the Virgin and the Angel'.[18] It was believed at
Hereford in the early nineteenth century that at some point the map
had been hidden under the floor of the cathedral's Audley Chapel, a
chantry beside the Lady Chapel; though there is no other evidence,
this is quite possible – the Lady Chapel for a long time housed the
library.[19]

Gough introduced the map to the antiquarian public with a full
description accompanied by an engraving of its representation of the

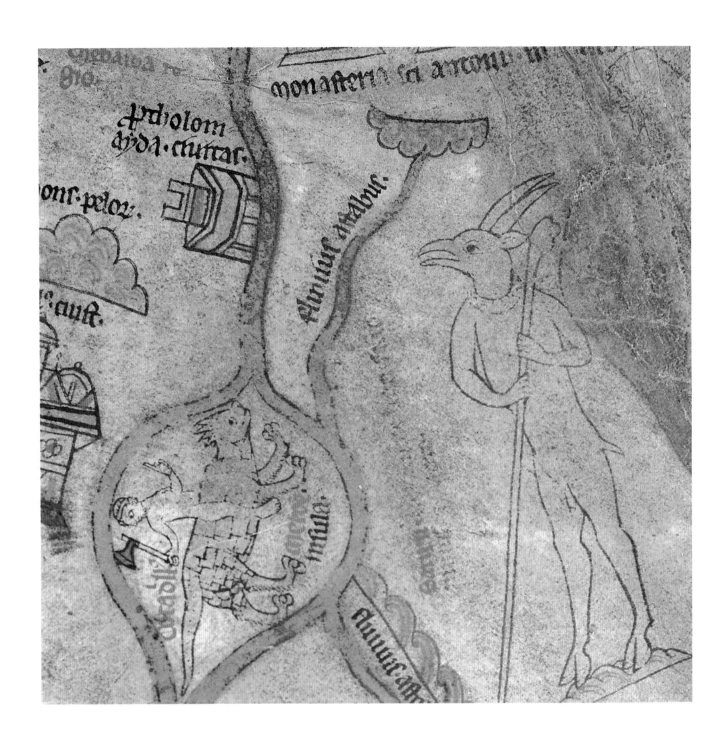

British Isles, and John Carter showed its case on the engraved title-page of his *Specimens of the ancient sculpture and paintings*, of which the first volume appeared also in 1780.[20] Even so, little importance was attached to the map at Hereford, and we have seen how lanterns had been piled against it in 1812. In 1831 a copy of it was made for the Royal Geographical Society; from this another copy was made in 1841 for the Bibliothèque du Roi in Paris and this in turn was reproduced by E.-F. Jomard in his *Les monuments de la géographie*. By this time it had been several times described and discussed by writers on medieval geography and maps. In 1855 it was sent to the British Museum to be cleaned

AN INSCRIPTION ERASED
The erasure of the inscription describing the satyr is probably deliberate, though there is some accidental damage in this part of the map. It has so far eluded attempts to read it. The satyr, horned, beaked and cloven-hoofed, is shown in upper Egypt because the fauns and satyrs tempted St Antony, founder of monasticism in the desert. On the left, on an island in the Nile, is a man riding a crocodile.

and repaired and since then, in the words of Bevan and Phillott, 'it has been treated with the most reverential care'. It had been moved from the library to the cathedral treasury in 1830, and was then displayed successively in the south choir aisle (from 1863) and the south transept (from the early twentieth century).

In 1867 a project was set on foot in Hereford to publish by subscription a new and improved facsimile of the map.[21] This had three important results. First, as a basis for the work, the map was in 1868, for the first time, photographed in four sections by Thomas Ladmore, professional photographer of Hereford. Second, the facsimile itself was published in 1872 after delays caused by the Franco-Prussian War; it was prepared in Bruges by a lithographer who also had a workshop in Paris, and one of those working on the map was killed at the battle of Sedan. Third, in 1873, a book was published to accompany the facsimile, written with verve and with great learning by W.L. Bevan, vicar of Hay-on-Wye, and H.W. Phillott, rector of Staunton-on-Wye; they describe the map in detail, identify the sources of its information and relate it to other medieval maps – a pioneer work of great importance. The map had been shown in June 1862 at the large special exhibition of works of art at the new museum at South Kensington, the future Victoria and Albert Museum, but it does not seem to have left Hereford again until the Second World War, when it was moved, for safety, first to the wine-cellar of Hampton Court, Herefordshire, then to a coal-mine at Bradford-on-Avon, Wiltshire. After the war, in 1948, the map again underwent conservation work at the British Museum,[22] and on its return to Hereford it was displayed in a new case, in the north choir aisle, provided at the expense of the Royal Geographical Society.

From November 1987 to March 1988 the map was exhibited at the

SIGNATURE OF JOHN NICHOLS
In the bottom right corner of the map is the signature 'John Nicolles' in an inexpert hand of the early seventeenth century. The ink is the same colour as the rings marking dappling on the horse, and one possibility is that John Nichols himself added these and then signed his work.

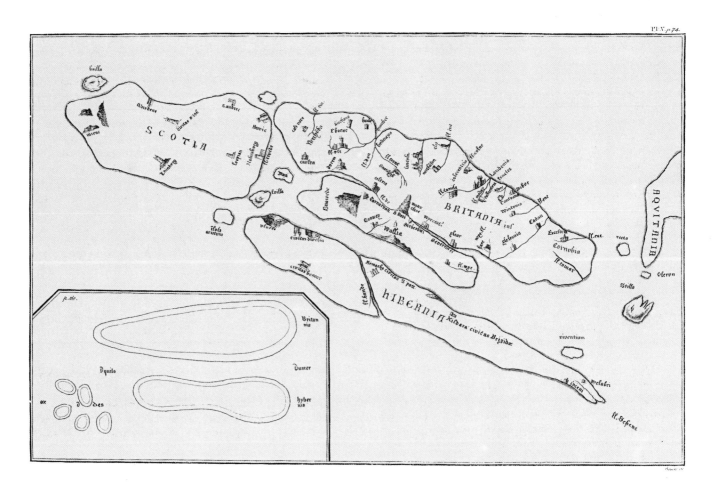

Royal Academy, London, at the important Age of Chivalry exhibition of English art of the thirteenth and fourteenth centuries. In November 1988, it was announced that the Dean and Chapter had instructed Messrs Sotheby and Co. to auction the map: the sale was planned for the following June, and it was hoped to raise a sum that would go far towards meeting the cathedral's long-term needs for the upkeep of the fabric. This aroused public protest and controversy. It was clear that the map was better known, more highly regarded and valued, than might reasonably have been thought. But apart from concern for the map itself, the proposal raised questions of the legal protection of the national heritage, the maintenance of ancient church buildings and the role of the Church of England as custodian of monuments and treasures, questions of wide interest that for some weeks were keenly debated in the national press. The proposed sale was postponed, and in 1989-90 the map was displayed in London at the British Library. In 1990 the Mappa Mundi Trust was set up, with a capital endowment from the National Heritage Memorial Fund, to take over the ownership and care of the map and of the chained library and other historic books at Hereford. A generous gift from Mr J. Paul Getty Jnr, KBE, enabled the Trust to put up a new building for their preservation and display.[23] This building, designed by Whitfield Partners, was opened in 1996.

EARLIEST REPRODUCTION FROM THE MAP, 1780
Richard Gough's British Topography, *published in 1780, is a bibliographical guide to sources of information on local history. It includes the first printed reference to the Hereford map. Gough recognised the map's importance – he devoted six pages to a full description, illustrated with this engraving of the British Isles as shown on the map. Inset, bottom left, is one version of the map of the British Isles that appears in copies of the description of Ireland by Gerald of Wales.*

British Library

NOTES

1 Miller's description of the map's colours (iv, p.4) is based on the facsimile of 1872. There are in fact no traces of gold on the crown and chain of the Emperor Augustus, nor on the golden fleece, which is coloured yellow with black flecks; the route of the Exodus is not white but olive-brown, with black hatching in some places.

2 That the words are addressed by the huntsman to the rider (as Jancey, p.4), not (as Bevan and Phillott, p.5; Miller, iv, p.7) by the rider to the huntsman, seems likely from the position and attitude of both figures and from the words' place in the drawing. The words may well have been a well-known call in hunting (as Crone 1965, p.448), though not specifically mentioned in a near-contemporary treatise (*La vénerie de Twiti*, ed. G. Tilander (Cynegetica, 2; Uppsala, 1956)), or possibly a personal motto (as Bevan and Phillott, p.5; Miller, iv, p.7; Denholm-Young, p.312).

3 Throughout the rest of this chapter I have drawn on my own observation of the map, and I am most grateful to the Trustees for making it possible for me to examine it by having it removed from its case – a complicated and lengthy operation. However, the time available was necessarily restricted, and I cannot claim to have made more than a simple preliminary examination of the map's surface; the map could not be lifted to reveal any note or drawing on the back or any significant pattern of pin-pricks. Much of the scholarly work done on the map has relied entirely on one or other of the facsimiles, and we could probably learn a great deal about the map's composition from a thorough physical and palaeographical examination of the map itself – to repeat a point I made over twenty years ago (Yates, p.171).

4 Bevan and Phillott, pp.2-3, Miller, p.6, and Crone 1954, p.3, take the name to be Richard of Holdingham 'and' of Sleaford, reading the conjunction as *e* (and). It is in fact clearly *o* (or) on the map, but the photograph in Bevan and Phillott, following p.xlvii, is either flawed or, more likely, has been retouched to disguise the small stain at this point, and shows the conjunction as *e*.

5 Clee Hill must be intended, though 'Clece' is an eccentric form (M. Gelling, *The place-names of Shropshire*, part 1 (English Place-Name Society; 1990), pp.82-7).

6 J. le Neve, *Fasti ecclesiae anglicanae 1066-1300. III: Lincoln*, ed. D.E. Greenway (London, 1977), p.20; archives of the Dean and Chapter of Canterbury, Sede Vacante Scrapbook I, p.142 (I am most grateful to Dr Nigel Ramsay for this information).

7 It is reassuring to find that this was the opinion of the late Professor Francis Wormald (Crone 1954, p.4). Reluctantly, I disagree with A.-D. von den Brincken, 'Monumental legends on medieval manuscript maps', *Imago Mundi*, 42 (1990), p.21, who suggests that the wording in red capitals around the edge of the parchment is in a different hand.

8 These are the garbled inscription at Ur, where the errors seem to include writing 'habet' for 'habre', probably simply an incorrect expansion of 'hab" on the draft; the reference in the Nile delta to the unfamiliar authority Artemidorus, which has been changed to the more familiar Isidore, but meaninglessly, as 'artim ysidorus'; the rectilinear pattern in Germany between the Danube tributaries and the Elbe and Weser. The note about Orosius above the horseman in the bottom right corner may also have been entered on the map by mistake, through misunderstanding the draft or other instructions (below, p.45).

9 This seems likely from the map's readings of 'habet' for 'habre', 'anim' for 'animal' (see notes 8, 10).

10 Others are the insertion of the *al* in 'animal' in the description of the *bonnacon* and of the *r* in 'Samarcan', Samarcand.

11 Bevan and Phillott, p.11.

12 Bailey, p.377.

13 This can be clearly seen in the photograph, made in 1868, of the lower left corner of the map in Bevan and Phillott, following p.xlvii, and on a much smaller woodcut of the whole map in *A handbook to Hereford Cathedral* (London, 1864), plate XI.

14 K. Edwards, *The English secular cathedrals in the middle ages* (2nd edn, Manchester, 1967), p.190.

15 *The history and antiquities of the city and cathedral-church of Hereford* (London, 1717), pp.64-5; J. Duncomb, *Collections towards the history and antiquities of the county of Hereford* (Hereford, 2 vols, 1804-12), i, p.587.

16 I owe this interesting suggestion to Mr Dominic Harbour. That this signature, faint but by no means obscure, seems not to have been noticed in print before, is a measure of how little the map itself has been examined by those who have written on it.

17 *History from marble, compiled in the reign of Charles II by Thomas Dingley, gent.*, ed. J.G. Nichols (Camden Society, old series vols 94, 97; 1867-8), i, pp.35-6, clx.

18 R. Gough, *British topography* (London, 2 vols, 1780), i, pp.71-6.

19 Bailey, p.377 and n.19.

20 Not, however, on the title-page of Carter's *The ancient architecture of England* (1795), as Miller, iv, p.4n, discovered; the mistake originated in Bevan and Phillott, p.11.

21 The progress of the project can be followed in reports in the *Transactions of the Woolhope Naturalists' Field Club*, in the volumes for 1867, pp.108-10; 1868, p.238; 1869, pp.152-4; 1870, pp.252-3; 1871-3, p.134.

22 An account of its cleaning and repair is in *Hereford Cathedral News*, 4 (Christmas 1948), pp.11-12.

23 *National Heritage Memorial Fund: Eleventh annual report 1990-1991* (London, 1991), p.1.

20

2 · The map and its relatives

MEDIEVAL AND ROMAN WORLD MAPS

How important is the map? There are, after all, over a thousand medieval world maps in existence – they are by far the most common type of map that survives from the middle ages, and probably by far the most common type that was produced. However, most of these maps were of the simplest kind, no more than diagrams of the main regions of the world. Some were zonal maps, a circle divided into horizontal bands representing the tropical, temperate and frigid zones. Others made a gesture to geographical outline, the so-called T-O maps – a circle divided by a T (sometimes a Y) to show the three continents. These might seem two analogous ways of dividing the world into regions, the climatic and the geographic, but in fact each starts from a different basis. The T-O map covers only the inhabited world, whereas the zonal map is the entire globe seen from one side. We should not be misled by the Hereford map into supposing that learned persons in the thirteenth century thought of the world as a flat disc with Jerusalem at its centre. Most envisaged it as a sphere, with its known inhabited part in the north, separated impassably from its unknown southern half by a belt of ocean and by the unbearable heat of the tropics. The circle of the T-O maps, the ocean-bounded circle of the Hereford map, would be viewed as a conventionalised form – a kind of projection – of what really occupied perhaps rather more than half the surface of a dome, the northern hemisphere of the world. The southern parts of Africa that we see on the Hereford map should be understood as lying within this northern hemisphere – Africa was not thought to extend to the equator, let alone beyond.

A few medieval maps show the fourth continent that was thought to lie in the southern hemisphere, beyond the equatorial ocean – but, indeed, many maps of the inhabited world were more elaborate and detailed than the schematic T-O maps. Most illustrate books that have geographical content or implications – encyclopaedias, lists of words and places, biblical commentaries. Sometimes the map appears in so many manuscript copies of a book that we can accept it as an integral part of it – it was either drawn by the author when he wrote the text or added by a copyist at so early a stage in the sequence of copying that it became effectively incorporated into the work. Thus a map of the world appears in fourteen surviving manuscripts of the commentary on the Apocalypse by Beatus of Liebana, manuscripts that date from the tenth century to the thirteenth; rectangular in shape, distinctively Spanish in artistic style and brightly coloured, they must derive from a common original which may well have illustrated the book when the author wrote it in the late eighth century. Another book consistently illustrated with a world map is the encyclopaedia by Lambert of Saint-Omer, and we know that this was the author's intention, for it was

DIAGRAMMATIC MAP OF THE
INHABITED WORLD, 11TH CENTURY
Many medieval maps of the inhabited world were no more than diagrams showing the relative positions of the three continents – a circle divided by a T or Y. Deriving ultimately from the same Roman sources, it can be seen as the Hereford map reduced to its bare essentials – the inhabited world shown in conventional form as a circle.
British Library, Royal MS. 6 C.1, f.108v

Opposite:
HIGDEN WORLD MAP, LATE 14TH
CENTURY
The universal history written in the mid-fourteenth century by Ranulf Higden, monk at Chester, was illustrated by a map entirely in the tradition of the world maps of thirteenth-century England. Higden's work was popular and often copied, and there are many versions of his map. Here England, bottom left, is given special status, coloured red and covered with cities, but Wales has become a separate island.
British Library, Royal MS. 14 C.IX, ff.1v-2

21

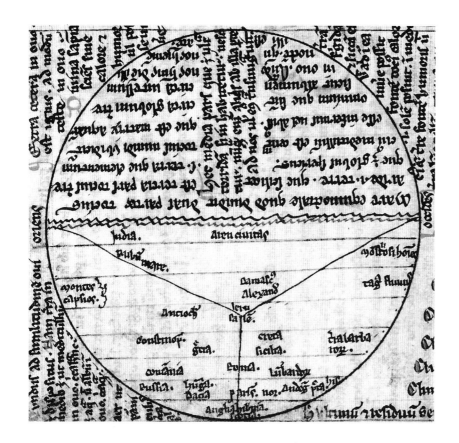

included in his autograph text, written in the early twelfth century. But
an individual copyist might, on his own initiative and using some other
source, change the map in the text he was copying or add a map to a text
hitherto unillustrated. The encyclopaedic works of Isidore of Seville,
written in the early seventh century, normally contain a more or less
simple T-O map, but one twelfth-century copy now at Munich has a
map showing cities, rivers and mountains that is visibly akin to the
Hereford map, though only 10 inches (26 centimetres) across. We must
be careful not to assume that the original author of a work had anything
to do with a map that appears only in a copy made centuries later.

Simple or elaborate, large or small, all these maps of the inhabited
world are related to each other and to the Hereford map. In the broad-
est terms, features of their outlines are common to all. They all belong
to a single, much ramified tradition which must go back to the Roman
period. We cannot suppose that people between the fifth century and
the twelfth themselves mapped any substantial part of the known
world. The techniques of surveying and of calculating geographical coordinates had fallen into disuse, the necessary administrative structure
no longer existed and, most important of all, the concept of mapping to
scale, of mapping from measurements, had been lost. It follows that
where any portion of a world map of this period has a recognisably
correct outline, this must go back to a Roman original, and that this
original was probably a measured and reasonably accurate map of the
world, showing coastal outlines, mountains, rivers, towns and boundaries of provinces. It need not surprise us that these recognisable

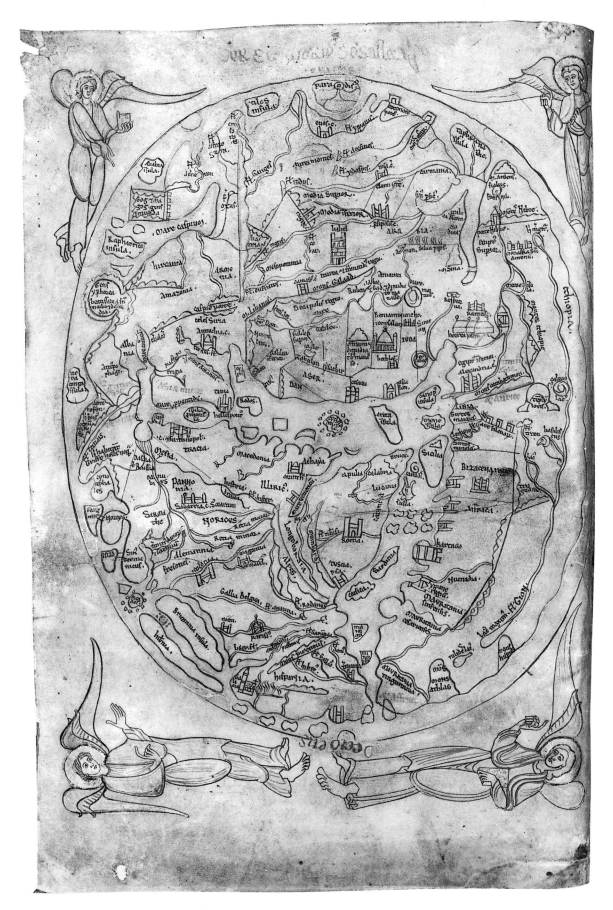

23

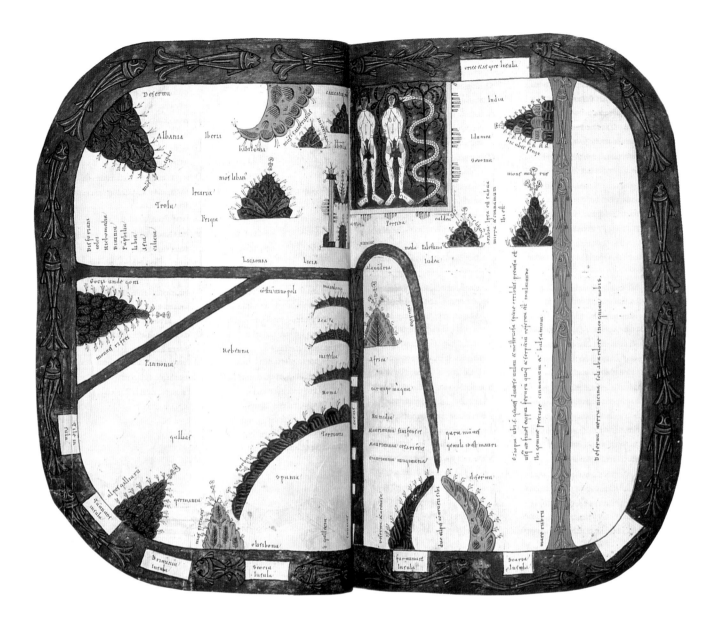

outlines have only occasionally survived the process of transmission –
on the Hereford map the triangular Sicily is practically the only
example. The outlines that we have are the outcome of many successive copies made by draftsmen who perhaps had little idea of the nature
of cartography and certainly no idea at all of the actual shape of any part
of the world. In their hands any error will all too easily have been
repeated and magnified – or the outline may even have been deliberately distorted, to force the map into a circle or other arbitrary frame.

Serious doubt has recently been cast on the Romans' cartographic
awareness – how far they really made and used maps.[1] However, it
seems inescapable that not later than the fourth century copies were in
existence of a world map – or of more than one – that was based probably on measured survey, possibly on geographical coordinates, and that
this map contained, at the very least, all the most accurate features that
can be detected in any surviving medieval world map. O.A.W. Dilke
suggests that the account of this map's creation in the Hereford map is

BEATUS WORLD MAP, EARLY 12TH
CENTURY
*When Beatus of Liebana wrote his
commentary on the Apocalypse in the eighth
century he may well have illustrated it with
the map which appears in varying form in
surviving later copies of the book. East is at
the top, with Adam, Eve and the serpent,
and the vertical brown line in the middle of
the map is the Mediterranean. On the right,
separated by the red band of the equatorial
ocean, a fourth continent is shown in the
southern hemisphere.*

British Library, Additional MS. 11695, ff.39v-40

24

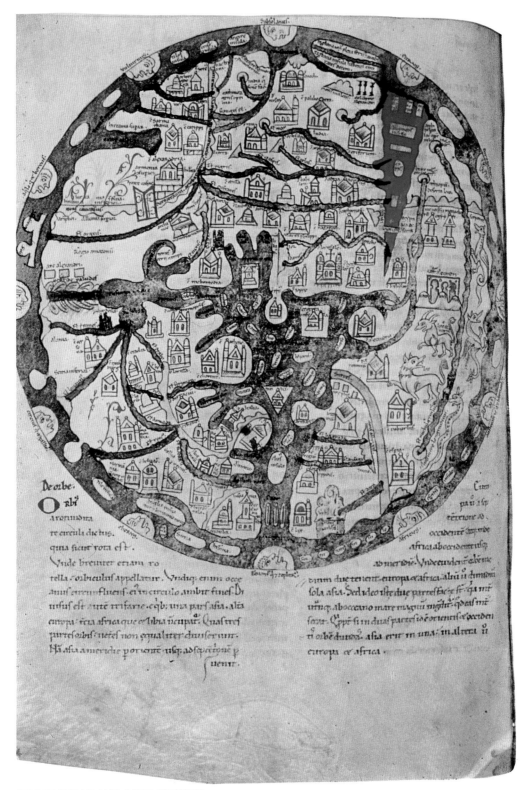

ISIDORE WORLD MAP, 12TH CENTURY

The world map usually drawn to illustrate the encyclopaedic works of Isidore of Seville was a simple T-O diagram. However, in this manuscript from northern France, the map is more elaborate, and there are interesting resemblances to the Hereford map. Thus the twelve winds appear around the edge, Noah's ark, Alexander's camp and the desert monasteries of Egypt are shown, and there is a series of exotic animals along the southern edge of Africa.

Munich, Bayerische Staatsbibliothek, Clm 10058, f.154v

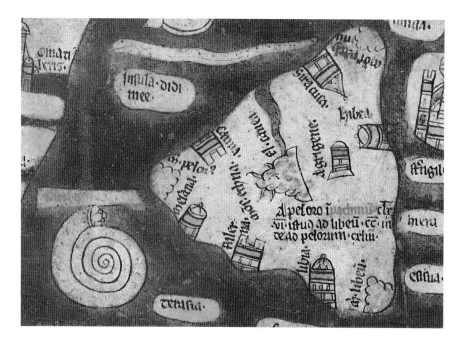

no more than slightly garbled. Two accounts from the late Roman period tell how in 44 BC Julius Caesar sent out four surveyors to measure the world – the three of the Hereford map (but Nicodemus, not Nichodoxus) plus Didymus who measured the western quarter. It was thirty-two years before they all finished, in the reign of Augustus. Their work will have served as the basis for a world map which – as recorded by the nearly contemporary Pliny – the emperor ordered Marcus Vipsanius Agrippa to set up for public display in a colonnade at Rome. However, Dr K. Brodersen argues that the evidence for Caesar's survey is too late, and the details too questionable, to be believed, and that what Agrippa produced was not a map but a written text, a geographical description with measurements.[2] It seems that for the Roman map of the world we have no firm evidence but its travestied medieval derivatives.

The phrase *mappa mundi*, literally cloth of the world, is unknown to classical Latin. It is first recorded in the early ninth century,[3] and it was what a world map was normally called in the twelfth and thirteenth centuries, whether drawn on cloth or not. In applying the phrase to the Hereford map we are calling it what Richard of Holdingham will have called it himself. However, the meaning of the phrase broadened to include other sorts of map – already in the late thirteenth century we find a chart of the Mediterranean referred to as a *mappa mundi*[4] – and from it English and some other languages derive their modern word for maps of all kinds. The full phrase occurs in English in the fourteenth and fifteenth centuries as *mappemounde*, and it still survives in French and Italian as *mappemonde*, *mappamondo*, meaning solely a medieval map of the world.

THE ENGLISH FAMILY OF WORLD MAPS

One medieval world map that seems closer than any other to the Roman original is the Cotton map, so called because it is among the manuscripts collected by Sir Robert Cotton (1571-1631) and now in the British Library. Compared with the Hereford map, the relative accuracy of its outlines of Jutland, the Frisian coast, Britain, Ireland, the Peloponnese and the Nile delta is striking. It dates probably from the second quarter of the eleventh century and it accompanies, suitably enough, a copy of Priscian's fifth-century Latin translation of the 'Periegesis', a description of the world written in Greek verse by an author named Dionysius. Map and text survive together only in this one manuscript, and it may have been the copyist's idea to unite them. The map is sometimes called the Anglo-Saxon map, and its spelling of the name it gives the Bretons – 'Suðbryttas', using a runic letter for *th* – shows that it was of English origin.

This is especially interesting. It may well be that in early medieval England there were more copies of the Roman world map, in fully detailed form, than in other areas of Europe – and better copies, with outlines less distorted, though people at the time could not have known this. Certainly in the twelfth and thirteenth centuries immensely detailed world maps developed as a genre which, though not exclusively English, came to be centred on England. However, the genre probably had its origins in France. In the early twelfth century a large world map, which included information on the peoples and the natural history of distant lands, was described in a little work by Hugh of Saint-Victor, perhaps a lecture written down by a pupil.[5] The twelfth-century Munich copy of Isidore of Seville's work, with elaborated map, was produced in northern France, and the so-called Jerome maps are in a manuscript from Tournai. In the church at Chalivoy-Milon, 25 miles (40 kilometres) from Bourges, a vast world map, 6 metres (20 feet) across, was painted in the mid-twelfth century. It was destroyed in 1885, and we know of it only through two short descriptions written when it was already much decayed, but from them it sounds very like one of this same group of maps.[6] However, both the Vercelli and the Ebstorf maps have possible English associations, and all other maps in the group were drawn in England. We know of others in England too that have failed to survive: there was a *mappa mundi* in Lincoln Cathedral library in the mid-twelfth century and another was among books that Bishop Hugh le Puiset left to Durham Cathedral Priory in 1195, Henry III had world maps painted on walls at Westminster Palace in the 1230s and at Winchester Castle in 1239, and Matthew Paris, a contemporary monk at St Albans, mentions one at Waltham Abbey and another by the otherwise unknown Robert *de Melekeleia*.[7] Of the late-twelfth or thirteenth-century world maps that survived to be known to us in detail, two are in books and three are separate large maps, like the Hereford map itself.

The earliest is the so-called Henry of Mainz map. The name arose through a misunderstanding; the map actually accompanies a copy of the early-twelfth-century encyclopaedia by Honorius of Autun. This copy was made probably at Durham in the late twelfth century and was then, with other works, given to the struggling Cistercian abbey at Sawley in Yorkshire. The map, naming Sawley in an inscription along the top, was added to the book perhaps as a kind of presentation frontispiece before it left Durham.[8] There is no reason to connect the map with the book's author but every reason to connect it with the map

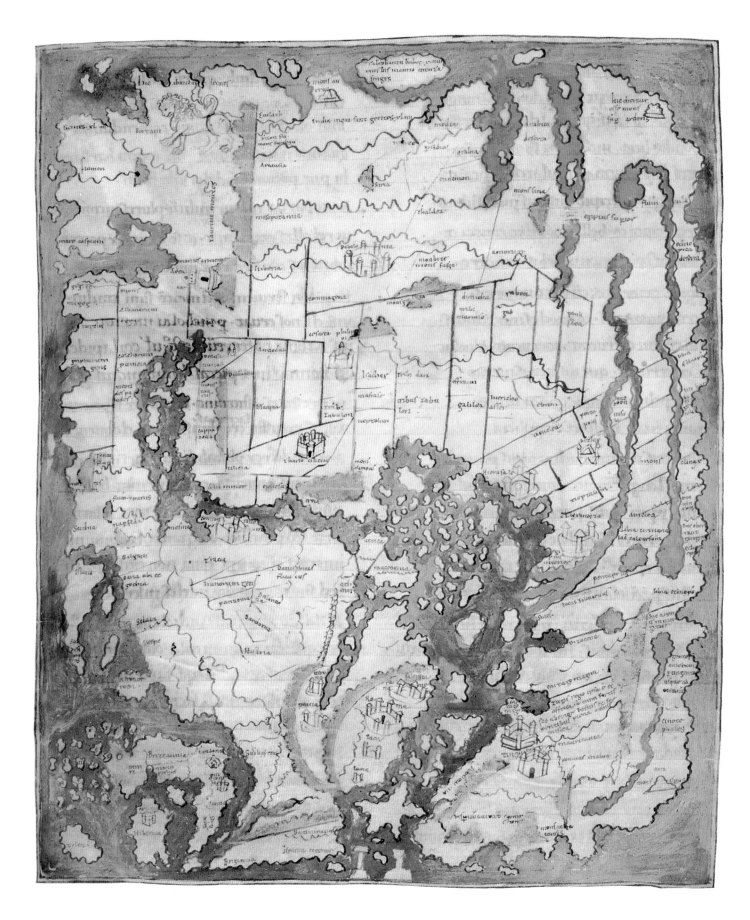

28

COTTON WORLD MAP, 11TH CENTURY
The Cotton map, of English origin, is closer in its outline than any other surviving medieval map to the Roman map of the world. The relative accuracy of the coastline of the North Sea and English Channel testifies to the quality of this Roman map. Though boundaries are prominent on the Cotton map it shows few rivers or towns.

British Library, Cotton MS. Tiberius B.V, f.56v

PSALTER WORLD MAP, MID-13TH CENTURY
This map was drawn on the first page of a psalter, produced in London in the early 1260s. Some 3½ inches (9 centimetres) across, it can be seen as a tiny picture of a world map. As on the Hereford map, Jerusalem is at the centre, but the terrestrial paradise at the top is within Asia, not an island, and we see only the faces of Adam and Eve.

British Library, Additional MS. 28681, f.9

that Le Puiset bequeathed to the cathedral priory. Although the map gives little general information and few pictures beyond the symbols for cities, we see links with the Hereford map in, for instance, the island of paradise at the top and Joseph's barns in Egypt.

More obviously related to the Hereford map is one which is an integral part of a psalter produced probably in London in the 1260s. It follows a prefatory series of scenes from the life of Christ that were added to the manuscript a little later, and is itself followed by the calendar which as usual precedes the text of the psalms.[9] It should perhaps be seen as a tiny picture of a large world map. Only 3½ inches (9 centimetres) across, it hardly has space for much information, but it includes fourteen pictures of monstrous races in southern Africa; like the Hereford map it is a circle centred on Jerusalem with twelve winds around the edge – little human faces blowing hard – and it has a markedly similar outer frame. Above, between two angels swinging censers, God raises the right hand in blessing and in the left hand holds

Left:

VERCELLI WORLD MAP, 13TH CENTURY
This map was probably brought back to Vercelli, in north Italy, by the bishop, Gualo Bicchieri, who was papal legate to England in 1216-18. Now much damaged and faded, it is sketchily produced and may have been meant as a draft for something like the Hereford map. Its general layout is the same, with east at the top, but the Mediterranean islands are drawn excessively large, so that the sea itself is reduced to a series of narrow channels between them.

From Y.Kamal, *Monumenta cartographica Africae et Aegypti* (privately printed; 5 vols, 1926-51), f.997; original in Vercelli, Archivio Capitolare

Right:

EBSTORF WORLD MAP, MID-13TH CENTURY
This map from the convent at Ebstorf, in north Germany, was destroyed in an air-raid in 1943 but replicas have been made from all available evidence. Some 10 feet (3 metres) square, it was the largest medieval world map that we know in detail. The head, hands and feet of Christ appear within the map's frame at the top, the bottom and each side. The portion missing from the upper right side of the map must have been removed deliberately, before the map came to light in 1830.

Ebstorf, Kloster Ebstorf

an orb that itself symbolises the world – it is in effect a minute T-O map. Below are two wyverns, two-legged dragons. On the back of the same leaf is a companion piece, a T-O map held before the figure of God with a list of provinces and cities in the space for each continent and again with two wyverns below. The purpose of these two maps is clear from their context: displaying the extent and variety of God's earthly creation they invite reflection on his omnipotence and are didactic in this spiritual sense – that they also convey geographical information was of secondary importance.

The most mysterious map of the group is in the cathedral library at Vercelli in north Italy. On a single piece of parchment and now badly damaged and faded, it was drawn as a near circle, probably slightly flattened at the sides, 33 inches (84 centimetres) from top to bottom. It has been variously assigned an Italian, a Spanish and a French origin, and has been dated either early or late in the thirteenth century as the Philip, king of France, whom it shows in north Africa may be either

Philip II (1180-1223) or Philip III (1270-85). In fact it is simplest to connect its presence at Vercelli with the scholarly bishop Gualo Bicchieri, who was papal legate to England in 1216-18 and who brought back with him a collection of manuscripts – among them an outstandingly important volume of Anglo-Saxon poetry. C.F. Capello suggests that the map was drawn by the canons of Saint-Victor at Paris, with whom the bishop made contact on his way home, but the picture of King Philip is not necessarily complimentary – he is shown riding an ostrich and wielding a scourge – and the map's apparent omission of the British Isles need not rule out an English origin, for in this part it is

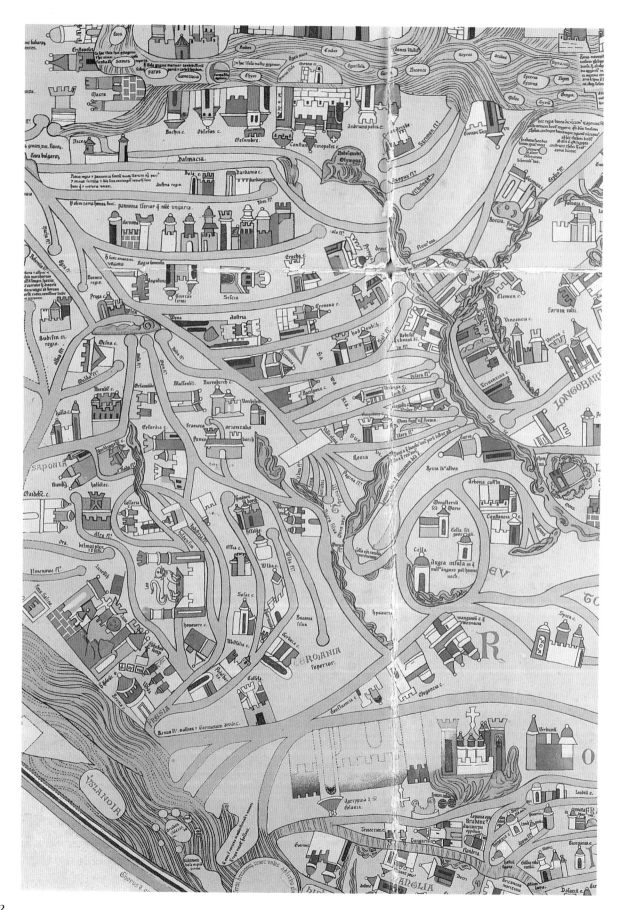

32

Left:

CENTRAL EUROPE ON THE EBSTORF WORLD MAP, MID-13TH CENTURY
For what it shows in Germany the Ebstorf map drew on sources quite different from the Hereford map, and there is evidence of local knowledge both at Ebstorf (Ebbekes storp) itself, lower left, where three small squares mark the five martyrs' graves at the convent, and at Reichenau, further to the right, where three separate monastic buildings are shown on the island near Konstanz and the source of the Rhine.

Ebstorf, Kloster Ebstorf

Right:

JERUSALEM ON THE EBSTORF WORLD MAP, MID-13TH CENTURY
On the Ebstorf map Jerusalem is shown as a walled square, with a picture of the Resurrection, whereas on the Hereford map it has circular walls with the Crucifixion above (see frontispiece). However, the two-humped camel on the left is probably related both to the Bactrian camel on the Hereford map and to the camel on Matthew Paris's map of Palestine.

Ebstorf, Kloster Ebstorf

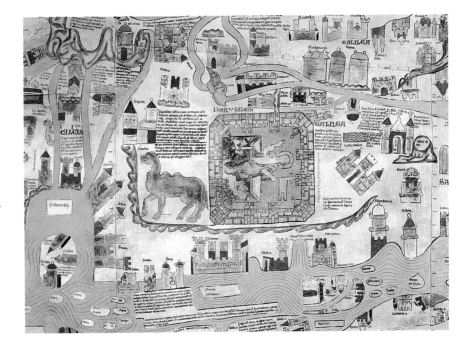

badly damaged. However, the map is probably the work of at least two hands, and its origin may well be complex. What is certain is that it is unfinished, and was produced far more roughly than the other world maps in the group. Its outlines were perhaps drawn by someone who did not thoroughly understand the map that was being copied – this could explain how the Mediterranean has been reduced to narrow channels between its many islands. The map may have been brought to Vercelli as a sketch from which a well executed world map might be made.[10]

The map most often compared with the Hereford map came from the convent at Ebstorf, near Lüneburg in north Germany, and from its local detail there is little doubt that it was made there. It has been convincingly dated 1239 from the political implications of what it shows. The prior of Ebstorf from 1223 to 1234 was named Gervase; he may or may not have been Gervase of Tilbury, of known English origins and connections, but the name was certainly rare in Germany while not uncommon in contemporary England. However, from its style and handwriting the map has been dated to the late thirteenth century, and one possibility is that it is a version made at Ebstorf of a map produced earlier elsewhere. The map itself was destroyed in an air-raid on Hanover in 1943, but four replicas have since been made with great care from all available evidence. It was much larger than the Hereford map, some 10 feet (3 metres) square and drawn on thirty pieces of parchment, but it had suffered some damage before it came to light in

DUCHY OF CORNWALL WORLD MAP, LATE 13TH CENTURY
This is all that survives of a world map probably contemporary with the Hereford map and slightly larger. The roundels at the bottom are the latter part of a series showing the ages of man. Unlike the Hereford map, sixteen, not twelve, winds were shown, heads blowing horns.

Duchy of Cornwall Office, Maps and Plans 1

1830 – parts of the northern edge of the map were lost and a rectangular portion in the south-east seems to have been cut out deliberately. The map was circular, with Jerusalem at its centre, and the twelve winds were marked with roundels, left blank, in the surrounding ocean. Also just within the circle of the map were the head, feet and hands of Christ – instead of bearing the world before him, as in the T-O map in the psalter, he is himself fully involved in it. Again the map displays an aspect of God's relationship with the world. The map was covered with pictures and inscriptions, just like the Hereford map; in the four corners were long extracts from books on geography and natural history and also a note on Julius Caesar's survey of the world, but without naming the surveyors.[11]

The one surviving corner of the Duchy of Cornwall map, the remaining map in the group, is also filled with text, but more closely related to the Hereford map: it names the same three surveyors of the quarters of the world and gives the number of seas, mountains, provinces, cities and so on that were covered by each. All we have of the map is a piece of its bottom right corner, some 2 feet square (62 by 53 centimetres). It is enough, however, to show that the map was circular, slightly larger than the Hereford map and similar in character – we see some of the animals and monstrous races of Africa, pictured and described, and three of what must have been sixteen winds, human heads blowing horns. The fragment survived in the duchy's archives as a wrapper for records of Ashridge College in Hertfordshire; this was a house of Bonhommes, priests living by the Augustinian rule, founded in 1283 under the patronage of Edmund, the king's brother. A book given to the college on its foundation is written by the same hand as the map, which may well have been made and presented at the same time.[12]

In all these world maps we see the map used as a vehicle for varied information arranged geographically. But they are alike not only in concept; they have enough detail in common to show that they are not independent productions – one way or another they are all related. On the other hand they are not simply variant versions of a single map – a glance shows that there are significant differences even in their outline, and in their detailed contents there are far more. To take a few random examples from the two major maps, the earthly paradise is an island on Hereford, part of the mainland on Ebstorf; Hereford shows the route of the Exodus, Ebstorf does not; in Palestine only a few of the coastal cities are the same on both, while Jerusalem appears as a circle on Hereford, as a square on Ebstorf; the *cynocephali* – the dog-headed people – are placed in Scandinavia on Hereford, in Ethiopia on Ebstorf. These examples could be multiplied many times. That the maps we know are so varied suggests strongly that there were once many others of which we know nothing – we have merely the few survivors.

This is borne out by contemporary regional maps that are related to the world maps. There are the amoeba-like maps of the British Isles in three copies of the description of Ireland by Gerald of Wales, and a map of Europe in a fourth.[13] There are the maps of Asia and of Palestine in a copy of works on biblical names by St Jerome (c.342-420) – they are known as the Jerome maps, but they do not appear in other manuscripts of the same works and they are in fact closely related to the Cotton map. Above all, there are the maps in the chronicles of Matthew Paris, monk

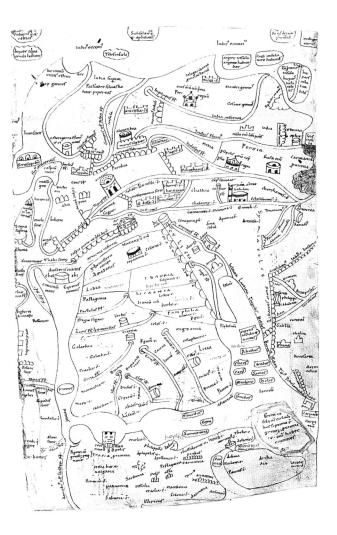

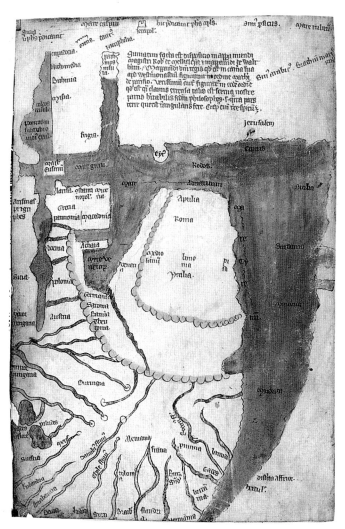

of St Albans, writer, artist and man of wide interests. One of his maps is often called a world map, but it is only part of one, showing Europe and the Mediterranean. The outlines of his four maps of Britain derive from world maps, probably from one with rivers and from another without,[14] and at least three of his four maps of Palestine are directly based on world maps, though the exact relationship has not been investigated. The world map may or may not have been a generally familiar object in thirteenth-century England, but it is clear that at the very least an educated person might expect to come across one from time to time.

Nor did the genre die out in the later middle ages. Ranulf Higden, who wrote successive versions of his immensely popular 'Polychronicon' from the 1330s to the 1360s, included a world map to illustrate its first book on natural history and geography – the other six books are a world history. A closely related map, with England much enlarged, was made as a wall map for Evesham Abbey in about 1390.[15] These maps are entirely in the tradition of the thirteenth-century world maps, and in this they differ from some other late-medieval English maps, which draw on the new, vastly more accurate, portolan charts. These developed as maps for navigation, first in the Mediterranean, then for the more distant places that Italian, Spanish and Portuguese ships visited –

Opposite:

GOUGH MAP OF BRITAIN, MID-14TH CENTURY

In its local detail this is one of the most remarkable maps from medieval Europe, but where, how and why it was drawn are all unknown. Its coastal outline in the south and south-east is taken from the relatively accurate portolan charts that were now being produced by Italian and Catalan map-makers, but the rest of the British coast was unknown to Mediterranean seamen and its outline still derives ultimately from the same Roman maps as the Hereford map.

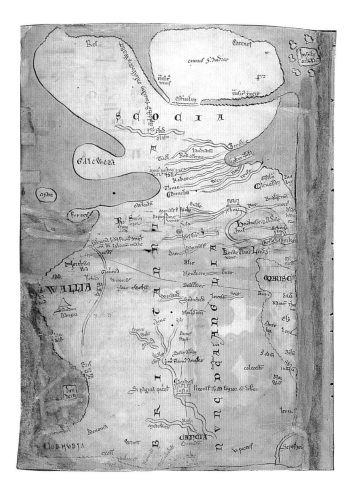

MAPS OF BRITAIN BY MATTHEW PARIS, MID-13TH CENTURY

Four maps of Britain by Matthew Paris survive. The coastal outline on each is taken from a world map on which its shape had not been distorted to fit within a circular frame. Despite this distortion on the Hereford map, its outline of Britain is in fact demonstrably akin to Matthew's maps.

Left:

'MAP D' OF BRITAIN BY MATTHEW PARIS

Matthew probably took this outline from a world map that did not show rivers. Those on the map will have been entered from personal knowledge – in some cases he gives the name of the river that passes through a town, like the Trent at Newark, but shows nothing more of its course.

British Library, Royal MS. 14 C.VII, f.5v

Right:

'MAP A' OF BRITAIN BY MATTHEW PARIS

On this, the most elaborate of Matthew's maps of Britain, the rivers as well as the coastal outline came from a world map and are related to those on the Hereford map. An example is the way a river cuts off Cornwall, here very nearly and on the Hereford map completely.

British Library, Cotton MS. Claudius D.VI, f.12v

36

Below:

MAP OF PALESTINE BY MATTHEW PARIS, MID-13TH CENTURY

There is a clear connection between Matthew's maps of Palestine – this is one of several – and the Hereford and other world maps. The large walled enclosure is Acre, the last town to remain in the crusaders' hands. Jerusalem is top right, a much smaller walled square, and above it are the Dead Sea and the River Jordan.

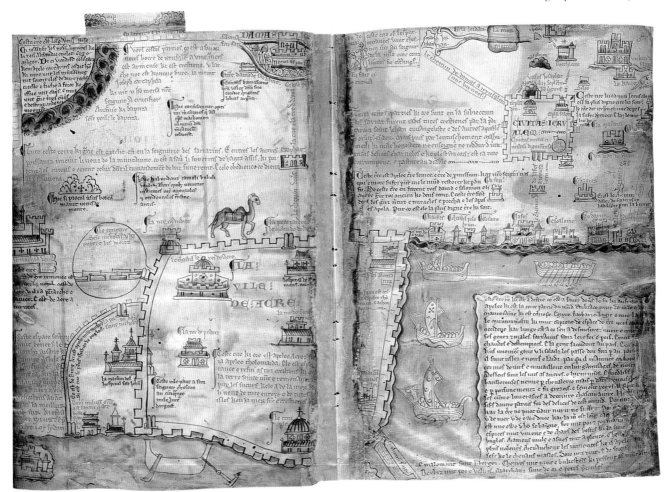

at first the Atlantic coast and the English Channel, much later the coast of west Africa. Their influence appears in the Aslake world map. This, like the Duchy of Cornwall map, is no more than a fragment used as a later binding, but it dates from the second half of the fourteenth century, and while it is closely related to the Duchy of Cornwall and Psalter maps it is from the portolan charts that it takes the places it names in north Africa, as well as the Canary Islands beyond.[16] This dual influence appears strikingly in the coastal outline of the mid-fourteenth-century Gough map of Britain, the most remarkable and most mysterious of all medieval English maps in the complexity of its internal detail; the south and south-east coast, the English Channel and the North Sea, are of a shape instantly recognisable and taken from the portolan charts, while the western and northern coasts, unknown to Italian navigators, still follow the outline of the older English tradition. By the fifteenth century the most advanced, most innovative maps of the world came not from England but from Italy.

The Hereford map, then, belongs to a whole group of maps which can be seen as a culminating – though sterile – development in a line of tradition going back to the Roman period. We probably know of only a tiny proportion of the maps created in this genre in the thirteenth century, when it seems to have been particularly fashionable in England. Of these maps, the Hereford map is now the largest and most detailed survivor – and, like the Ebstorf map, it tells us a great deal about the form and content of those we have never known. It is an invaluable record and relic of an interesting stage of cartographic awareness and development.

NOTES

1 Most comprehensively by K. Brodersen, *Terra cognita: Studien zur römischen Raumerfassung* (Spudasmata, vol.59; Hildesheim, 1995).

2 Harley and Woodward, i, pp.205-9; Brodersen, op.cit., pp.261-87.

3 P. Gautier Dalché, *La 'Descriptio mappe mundi' de Hugues de Saint-Victor* (Paris, 1988), pp.92-3.

4 Guillaume de Nangis, 'Gesta sanctae memoriae Ludovici', in *Recueil des historiens des Gaules et de la France*, ed. J. Naudet and P. Daunou (Paris, 24 vols, 1738-1904), xx, p.444.

5 Gautier Dalché, op. cit.

6 M. Kupfer, 'The lost *mappamundi* at Chalivoy-Milon', *Speculum*, lxvi (1991), pp.540-71.

7 See pp.34-5; Morgan, p.196; B. Meehan, 'Durham twelfth-century manuscripts in Cistercian houses', in *Anglo-Norman Durham*, ed. D. Rollason, M. Harvey and M. Prestwich (Woodbridge, 1994), p. 446.

8 Gautier Dalché, op. cit., p.183n; Meehan, op. cit., pp.440, 445-6. In my opinion the inscription naming Sawley is in the same hand as the rest of the page.

9 Morgan, pp.82-5.

10 C.F. Capello, *Il mappamondo medioevale di Vercelli (1191-1218?)* (University of Turin, Memorie e Studi Geografici, no.10; 1976), especially pp.5-18, 117-25.

11 B. Hahn-Woernle, *Die Ebstorfer Weltkarte* (Ebstorf, [1988]), is a general, well illustrated account of the Ebstorf map. More detailed studies are in *Ein Weltbild vor Columbus*, ed. H. Kugler and E. Michael (Weinheim, 1991), among them the important work on its date by A. Wolf (pp.54-119), R. Kroos (pp.223-44) and H. Appuhn (pp.245-59). Of Wolf's work there is an abridged version in English, 'News on the Ebstorf world map: date, origins, authorship', in *Géographie du monde au moyen âge et à la renaissance*, ed. M. Pelletier (Ministère de l'Éducation Nationale, Comité des Travaux Historiques et Scientifiques, Mémoires de la Section de Géographie, no.15; Paris, 1989), pp.51-68.

12 G. Haslam, 'The duchy of Cornwall map fragment', in *Géographie du monde*, pp.33-44.

13 P.D.A. Harvey, 'Matthew Paris's maps of Britain', in *Thirteenth century England: IV*, ed. P.R. Coss and S.D. Lloyd (Woodbridge, 1992), p.114n. It is interesting, but probably not significant, that one of these maps of the British Isles is in an early-thirteenth-century manuscript that belonged to the Hereford vicars-choral; it is now National Library of Ireland MS. 700 (N.R. Ker, *Medieval libraries of Great Britain* (Royal Historical Society; 2nd edn, 1964), p.99).

14 ibid., pp.114-16.

15 P. Barber, 'The Evesham world map: a late medieval English view of God and the world', *Imago Mundi*, 47 (1995), pp.13-33.

16 P. Barber and M.P. Brown, 'The Aslake world map', *Imago Mundi*, 44 (1992), pp.24-44.

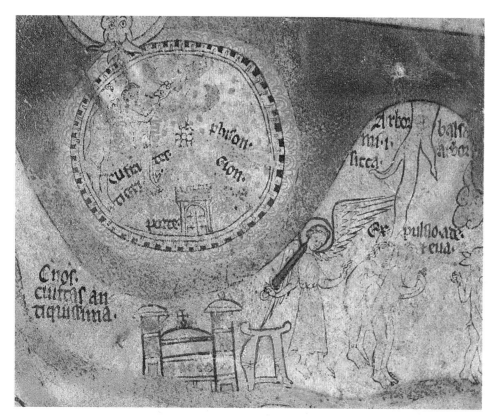

THE TERRESTRIAL PARADISE

The Garden of Eden is an island with a wall around it at the eastern end of the inhabited world. Its four rivers include the Tigris and Euphrates, and as these also appear on the map in Mesopotamia we realise that the map's geographical location of paradise need not be understood literally. On the right is the expulsion of Adam and Eve, and above them is the balsam, the dry tree (arbor sicca), from one of whose twigs grew the tree used for the Cross.

NOAH'S ARK

Noah's ark is shown where it came to rest in the mountains of Armenia. Below, beyond the mountains and the city of Malatya (Metima), is the tigolopes, a semi-human creature with webbed feet, holding what may be a thyrsus, a staff tipped with a pine-cone ornament and associated with Bacchus.

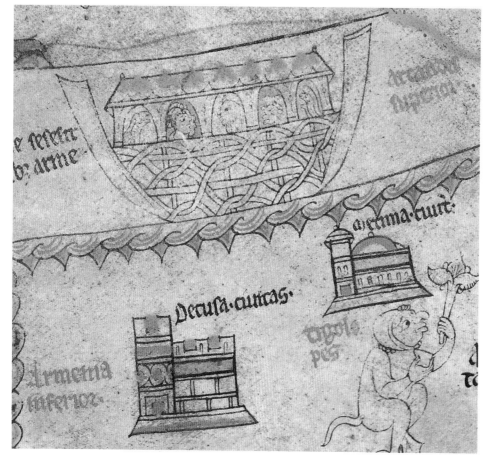

3 · The map and its sources

THE BIBLE AND EARLY CHURCH HISTORY

The Bible contributes less than we might expect to what is shown on the Hereford map. Certainly the events of, in particular, the Gospels were acted out in too small a region to allow much detail to appear. Yet Palestine is disproportionately large on the map and is far from over-crowded – there is more blank space there than in most of Europe or India. It lies, however, in the part of the map – Europe and the lands around the Mediterranean – that shows rivers, mountains and many towns, but not much else. Places from the Old and New Testaments appear in Palestine but not what happened there, though the picture at Bethlehem incorporates what may be a manger and the well of the oath is marked near the town to which it gave its name, Beersheba. Several of the twelve tribes are named; they all appear on some later medieval maps of Palestine. But the area is dominated by the circular, walled Jerusalem and by the space created by its surrounds – Mount Ephraim, the Mount of Olives, the Valley of Jehoshaphat – and here we see the Crucifixion, the only event from the Gospels shown on the map. The Ebstorf map shows instead the Resurrection.

Beyond Palestine the events of Genesis and Exodus are the only part

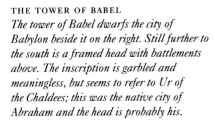

THE TOWER OF BABEL
The tower of Babel dwarfs the city of Babylon beside it on the right. Still further to the south is a framed head with battlements above. The inscription is garbled and meaningless, but seems to refer to Ur of the Chaldees; this was the native city of Abraham and the head is probably his.

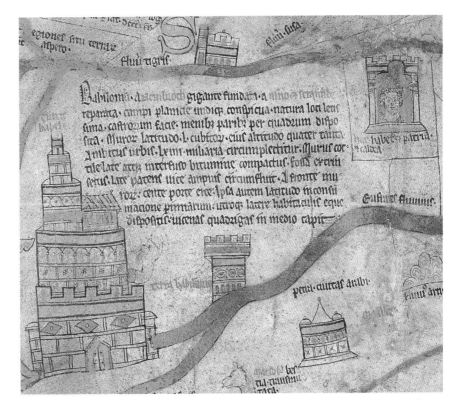

of the Bible to be well represented – we find the same pattern on the Ebstorf map. The terrestrial paradise and the expulsion of Adam and Eve appear in the extreme east of the map, Noah's ark where it came to rest in the mountains of Armenia. The tower of Babel dominates the regions between Palestine and India – it dwarfs the city of Babylon beside it – and to its south-east a framed head with battlements above is probably Abraham; the garbled inscription must have been mis-copied, but refers to his native city, Ur of the Chaldees.[1] The sites of Sodom and Gomorrah are named in the Dead Sea, and Lot's wife, turned to salt, stands beside – 'very forlorn-looking' say Bevan and Phillott. In Egypt are Joseph's barns, and the route of the Exodus starts nearby at Rameses. It crosses the Red Sea, and just beyond is Mount Sinai, where Moses is receiving the tablets of the law from the hand of God; the head of Moses has horns, following the Vulgate's rendering of the radiance that came upon his face, and below are the Israelites wor-shipping the golden calf. The route then winds and turns to show the wanderings in the desert before it crosses the Arnon and the Jordan and ends at Jericho. Other surviving world maps in the English group show the places on the path of the Exodus but not the actual line of the route; however, this appears on at least two earlier maps more distantly related – a reminder that the map's closest surviving relatives are no more than a few specimens of what may well have been a large as well as varied family of maps.

The only other book of the Bible that may have been systematically used for the map is the Acts of the Apostles; towns in Asia Minor and Macedonia, even Reggio and Pozzuoli in Italy, have perhaps been entered from the journeys of St Paul, though as they are dispersed among other places we cannot be sure of this. Of early Christian history we are offered no more than a few fragments. In front of Moses on Mount Sinai is probably the grave of St Catherine of Alexandria, a slab marked with a cross. Far up the Nile two substantial buildings are identified as the monasteries of St Antony (died 356), the desert father venerated as the founder of monasticism, and nearby a large figure in gown and cowl is named Zosimas, a fifth-century monk who retired to the desert. Finally, at Hippo in north Africa St Augustine (354-430) is shown, framed like Abraham at Ur but with spires instead of battle-ments above. All this amounts to only a tiny part of the information on the map; instruction in biblical and Christian history cannot have been the compiler's chief concern.

GEOGRAPHICAL WRITERS

The bulk of the general information on the Hereford map comes from nine classical and later Latin authors, most of them writers on geogra-phy. The sources of all the inscriptions – and thus the contribution of each author – were painstakingly identified by Bevan and Phillott in 1873 and revised in 1896 by Konrad Miller, who linked them to what appears on other maps. Five of the authors are in fact identified on the map itself, four of them in a few of the inscriptions that actually name their source. Thus the description of the rhinoceros comes specifically from Solinus, and the description of the unicorn not only names its

THE EXODUS
The Exodus is the part of the Bible that appears most fully on the map. Starting from Rameses, the treasure city built for Pharaoh by the children of Israel, the route crosses the Red Sea to Mount Sinai. There Moses receives the tablets of the law from the hand of God and – a little further along the route – the Israelites worship the golden calf. The route passes the phoenix, of which only one exists but which lives for five hundred years, before a series of loops marks the wanderings in the desert. It passes the Dead Sea, with the ruins of Sodom and Gomorrah, crosses the River Arnon and passes Lot's wife, turned to salt. Finally it crosses the River Jordan and reaches Jericho.

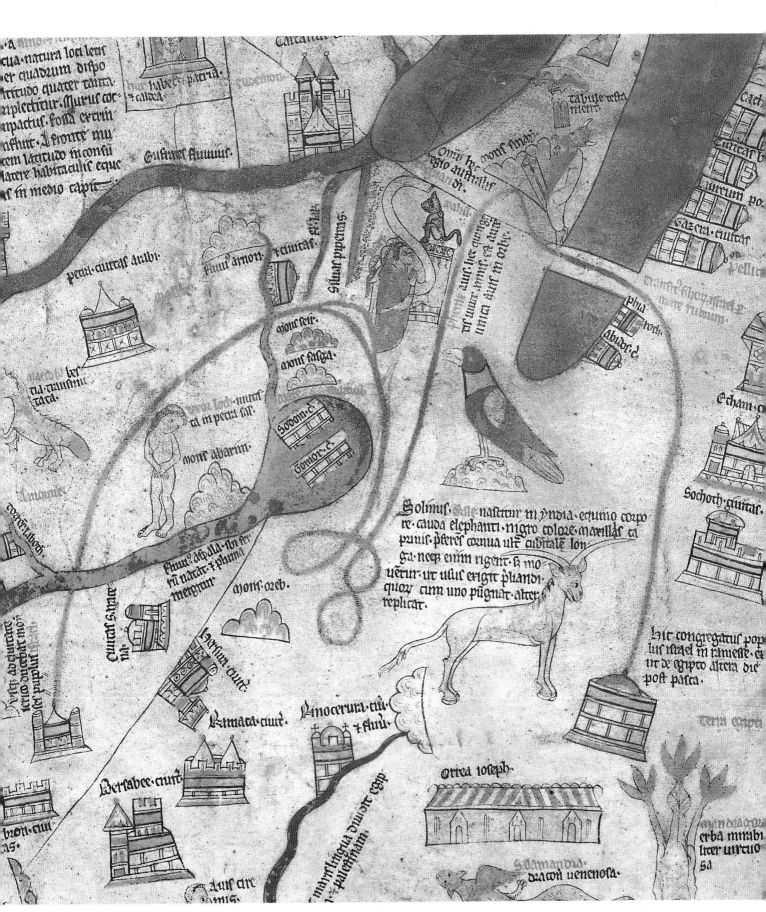

43

ST AUGUSTINE
*St Augustine (354-430), doctor of the
Church and one of the greatest early
Christian writers, appears on the map at his
see, Hippo Regius near Bône in Algeria.
Mostly, however, the map ignores the early
history of the Church.*

THE NILE DELTA
*At its mouth the Nile is shown dividing
between two branches and at the point where
the lower enters the sea a delta formation is
drawn. However, the island between the
branches is described as part of the delta,
where according to Artemidorus there were
250 towns. On the sea by the lower branch is
Alexandria, shown with its famous
lighthouse.*

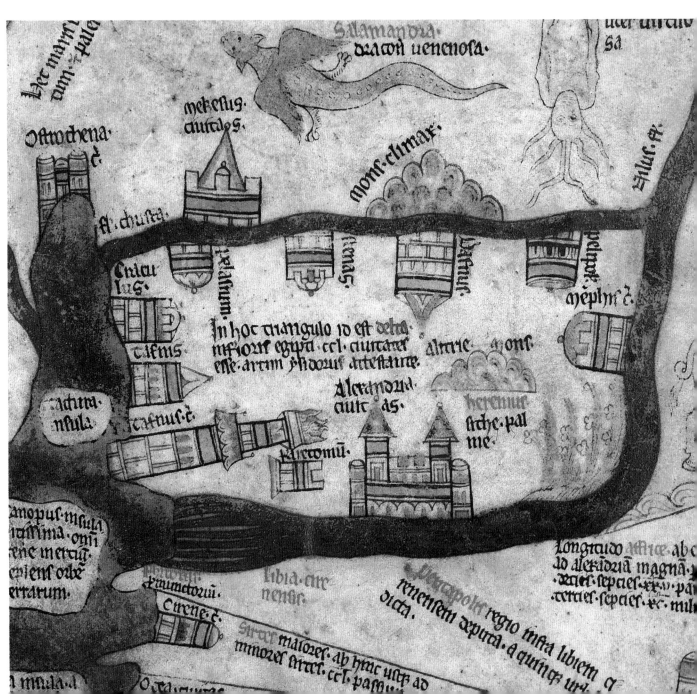

THE *EALE*
The eale, *the inscription tells us, lives in India – though in fact the map places it in Egypt and the long building below is labelled Joseph's barns, the name given to the pyramids. It has the body of a horse, the tail of an elephant and the jaw of a goat, and its horns, a cubit long, are flexible so that they can be used simultaneously in attack and defence. This is one of the few inscriptions that names its source, the third-century author Solinus.*

author, Isidore, but gives the full reference to his 'Etymologies', book 12, chapter 2. Aethicus and Martianus are similarly named on the map, and outside it, just above the horseman in the bottom right corner, is an inscription naming Orosius: 'Orosius's account of the *Ornesta* of the world, as shown within'. *Ornesta*, correctly *Ormista*, is an acronym used for Orosius's book, 'Orosii mundi istoria' (History of the world by Orosius). The inscription is seen by Bevan and Phillott as the title to the whole map, by Miller as an acknowledgment of a principal source, parallel to the reference to Augustus's survey in the opposite corner.[2] Neither explanation really fits either its position on the map or its wording – it reads rather like an ill-informed note from the fly-leaf of a manuscript containing Orosius's work, and its appearance here is unaccountable.

Orosius was, however, one of the authors most closely connected with the contents of the map. This is not to say that he was himself an original authority. The five writers named were no more than compilers from earlier works – thus Solinus took almost all he wrote from the works of Pliny the elder and Pomponius Mela in the first century AD, two hundred years before him, and they in turn will have used earlier sources. The *essedones*, who ate their parents' corpses, are described by Solinus and Pomponius Mela – but also by Herodotus in the fifth century BC. One inscription on the map reveals an authority behind its actual unnamed source, Martianus: in giving the number of towns in the Nile delta it refers back to Artemidorus, a Greek writer of about 100 BC. What these later writers offered were eclectic compilations from a general corpus of information on geography and natural history that originated in the classical period. Nor should we suppose that the map's author himself read these books and constructed the map from what he read – the appearance of so much of the same material on related thirteenth-century maps makes it clear that he did not. He may well have revised or added to what he found on his models by referring back to these sources, but any significantly original contribution that he may have made to the information on the map is not to be found in what came from these geographical writers.

Gaius Julius Solinus, of whom we know only that he wrote probably in the early third century, is the earliest author that the map uses directly. He called his book 'Collections of memorable matters' and from it the map derives much of its information on the monstrous races and strange customs of the most distant parts of the world. Of Paul Orosius we know rather more – he was a pupil of St Augustine at Hippo, having fled from Spain before the barbarian invasion. His 'Histories' were probably completed in 418, and from them the map may have drawn much purely geographical information, including rivers and mountains in Asia and Africa. However, Miller argues that this is to see things the wrong way round – rather, Orosius himself had access to an ancestor of the map and took his own geographical information from it. The map's only inscription that is certainly taken from Orosius is its lengthy account of Babylon. The encyclopaedic work of Martianus Capella, known as the 'Marriage of Mercury and Philology', also dates from the early fifth century, and from this the map takes three of its inscriptions – in the Nile delta, Persia and north-east Asia. From the work ascribed to Aethicus of Istria, which probably dates from the

Hic habitant griffe homines ne quissimi·nam inter cetera facinora·eciam de aritbz hostium suorum·tegumenta sibi·t eqis suis faciunt·

Longitudo ... ab ostio meotidis cp ad gadicani· ... aurtii·tercier quar·xxvii·passuu·vniusur autre arcutt· meotu lacu· cencier·quindgies·septier·nonaginta mi· Cun tpa meotide· cencier·quinquagier·xxxiii·nona suum·igungari·

sclau·

mid-eighth century, the map takes much of its account of the peoples of northern Asia. The map's connection with the work of Isidore, bishop of Seville from about 600 to 636, is clear but imprecise – thus, for instance, the account of each wind on the map can be found in one or other of his two relevant books, the 'Etymologies' and 'Nature of things', but is differently phrased. Indeed, besides identifiable quotations from all these five authors there is much on the map that is clearly related to writings of one or more of them but without following their exact wording. Much of what the map displays is from the general fund of information on geography and natural history that was current in the late classical and early medieval periods.

The other four authors identifiably used for the map made more distinctive contributions. It was from St Jerome's life of St Paul – the fourth-century hermit, not the apostle – that the map takes not only the monasteries of St Antony but also the nearby faun and satyr that he met in the desert. The eighth-century 'Acts of the Lombards' by Paul the Deacon is the source for the reference on Jutland to the seven sleepers, believed to be Romans from their dress, and also for the two heads marking whirlpools at the north and south ends of Britain – that both are named *Suilla* suggests confusion between the Scilly Isles and Scylla

THE *GRISTE*
These people, in the extreme north-east of Europe, made coverings for their horses and themselves from the skins of their enemies – an account which combines the descriptions of the griphe *by Aethicus and of the* geloni *by Solinus. To the left are the* cynocephali, *the dog-headed people, who appear on other maps in India or Ethiopia – the Hereford map follows Aethicus in placing them in Scandinavia.*

46

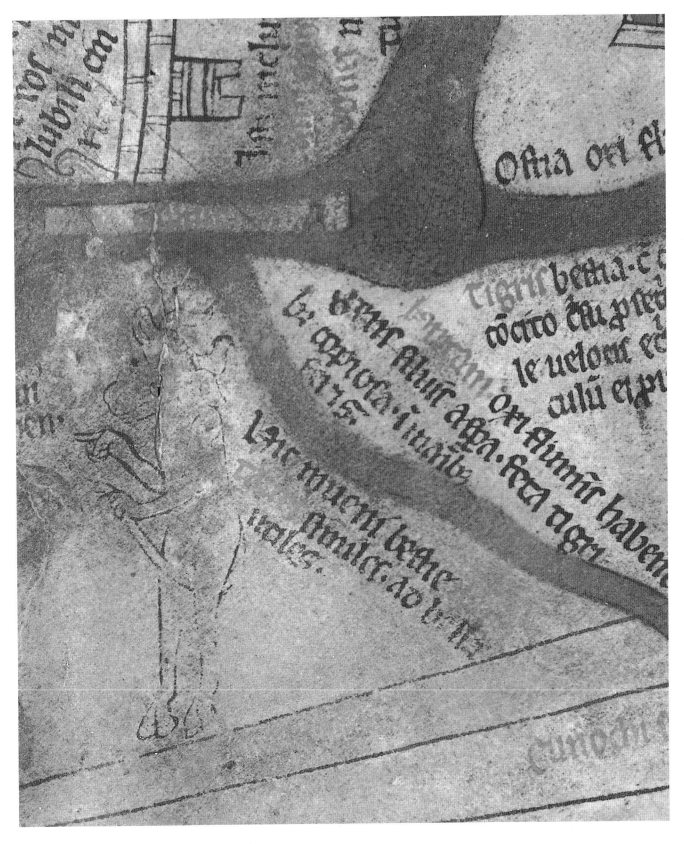

THE CREATURES LIKE MINOTAURS

'Here I found beasts like minotaurs, useful for war.' This inscription – the map's only use of the first person singular – is based on the eighth-century work attributed to Aethicus. The creature, partly human, partly bovine, holds its tail with one hand while making a conversational point with the other.

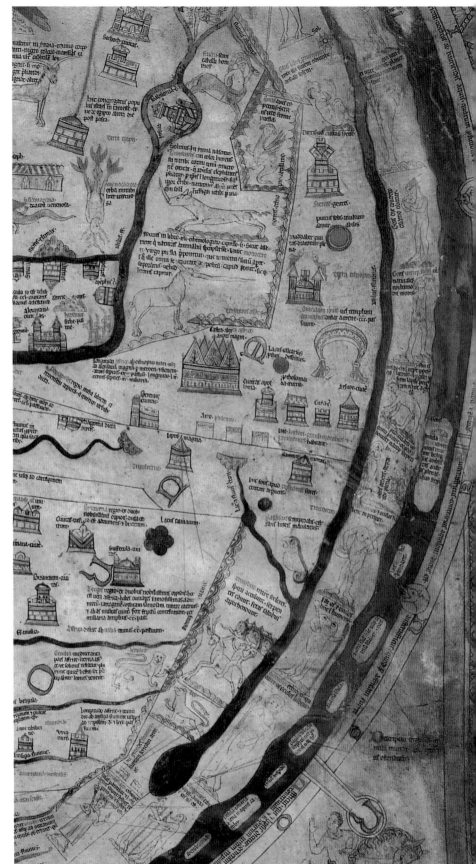

Right:

THE PEOPLES OF SOUTHERN AFRICA
Between the upper Nile and the ocean, the map shows a series of ten strange races, mostly of peculiar physique. They include the people with only one leg and one eye (second down), those who cannot open their mouth so have to take nourishment through a straw (third down), those who walk on all fours (fifth down) and two peoples with their face in their chest (seventh and eight down).

Below:

NORTH-WEST AFRICA
On the right are the northernmost of the Fortunate Isles, entered from the legends of St Brendan's voyages, mentioned in the inscription. Above is Mount Atlas, where at night there are lights and the sounds of cymbals and Bacchic song, and further north along the coast is Gibraltar (Mons Calpel), placed in Africa instead of Europe. Correctly, this was one of the pillars of Hercules, but they are shown, literally as pillars, on the island of Cadiz, at the mouth of the Mediterranean.

and Charybdis in the Straits of Messina. The islands of birds and of rams between Scotland and Ireland and the six Fortunate Isles, south of Gibraltar, come from the legend of St Brendan's voyages, as recorded in the tenth or eleventh century, while the River Eider, boundary between the Danes and Saxons, and notes of the Frisians and Slavs are taken, more reliably, from the eleventh-century 'History of the church of Hamburg' by Adam of Bremen.

While there has been much learned investigation into the literary and geographical sources of the Hereford and other world maps and into the links between them, there has been no research at all into the pictures that accompany so many of the inscriptions. That here too there are links can hardly be doubted. To take a single, simple example, the two-humped camel of the Ebstorf map is probably related to the two-humped camel at Hereford, and is almost certainly related to the camel on Matthew Paris's maps of Palestine – both are walking just outside Jerusalem. On the other hand the single-humped camel of the Vercelli map is clearly a different creature altogether. The

Below:

THE ADRIATIC AND THE PO VALLEY
How far the towns on medieval world maps derive, like the outlines, from Roman originals is uncertain. Some, however, must have been added, like Venice, a later foundation, shown as a prominent island in the Adriatic but sadly misplaced – it is further south than Durazzo and nearly opposite Delphi (Delos, upper right), where a man's face marks the oracle of Apollo. The mountains at the bottom are the Alps, those on the right the Apennines – even the Po valley is not crowded with towns.

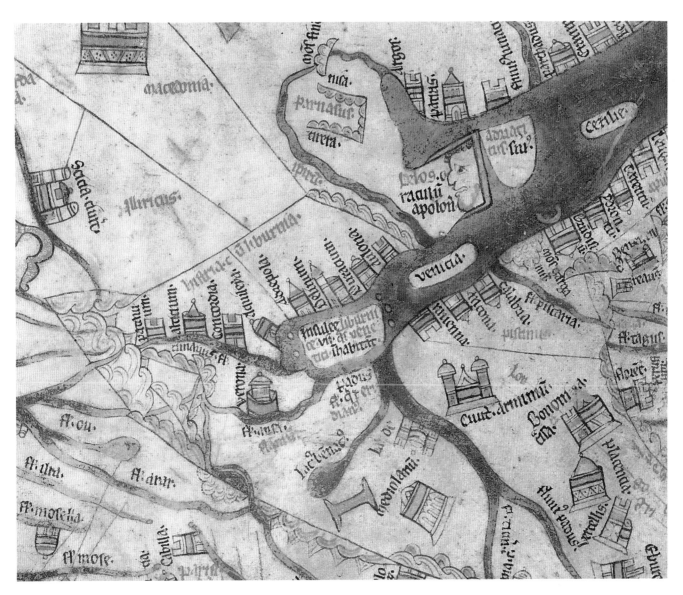

interconnections of these pictures, to say nothing of their sources, could well teach us much about the genesis of this group of world maps and of the relationships between them.

ITINERARIES

The Antonine Itinerary was first compiled probably in the early third century AD; it consists of many detailed lists of places along routes through the whole Roman empire – thus there are some fifteen routes through Britain. Bevan and Phillott name it among the sources used for the Hereford map and other medieval world maps, linking it with the map's place-names in north Africa and elsewhere. Miller, however, rejects any connection, arguing that the simple posting-stations and notes of distances characteristic of the itineraries simply do not appear on the maps; inevitably they show some of the places named in the itineraries, but we should see no direct textual link. In two discussions of the Hereford map in 1954 and 1965 G.R. Crone took the opposite view. He saw the map as a mass of itineraries with many or most of its towns entered from lists of places along routes – the route of the Exodus, the travels of St Paul, routes through the Roman empire from the Antonine and related itineraries, routes followed by medieval pilgrims and traders.

Behind these two opposed views lie entirely different assumptions about what was shown on the Hereford map's early ancestors. Miller, as we have seen, thought it more likely that Orosius took his information about rivers and mountains from an earlier map than that later maps drew substantially on Orosius as a source. In fact he went farther than this, suggesting that much of the general information on the Hereford map and its relatives had already been entered on maps in antiquity; the thirteenth-century map-makers may well have edited and revised, corrected and added to this material from literary sources, but they started with a great deal of information that was on the earlier maps they copied. Crone on the other hand assumed that even towns did not appear on world maps in any number until, say, the twelfth century; what came down from antiquity were little more than skeletal physical maps with provincial boundaries, which were then used as a framework to contain an encyclopaedic mass of information drawn from whatever sources were available. Both are extreme views. A sensible guess – it is no more – is that our map's Roman archetype showed coastal outlines, rivers, mountains, provincial boundaries and also a significant number of towns, but nothing else. There seems no reason why towns should have been left out, the more so as they appear in Roman itineraries, both written and graphic. On this assumption the Cotton map, despite its good coastal outline, would be no more than a summary sketch with many omissions, while the Jerome maps would offer a reasonably complete picture, however much distorted, of what appeared on the maps of imperial Rome – we may note that although they contain no encyclopaedic information they still have no awkward blank spaces, even in central Asia. There would be no need for a massive introduction of towns into the map at any later period – many would be there already.

But the Hereford map shows many towns that did not exist in the Roman period – Venice, Compostela, Dublin, Bremen are examples. Whether or not we accept Crone's view of the use of ancient itineraries in the map, his demonstration of the use of twelfth- and thirteenth-century itineraries is convincing and interesting. Lists of successive places along routes were compiled in the middle ages as in antiquity, and Crone finds traces of them on the map, particularly in France. Here the name of a mountain, *Recordanorum*, recalls the *voie regordane* of earlier Provençal poetry, a route to Provence from the north of France. Other sequences of names in France show commercial routes to the fairs of Champagne and Flanders and from Bordeaux to the Mediterranean, and pilgrims' routes to St James of Compostela. Crone sees a further trade route eastwards from Cologne, presumably through Gandersheim and Halberstadt to Magdeburg on the Elbe, but if it is there it has gone awry for the few towns entered in Germany seem all but arbitrarily placed; the map's line of towns along the River Weser is

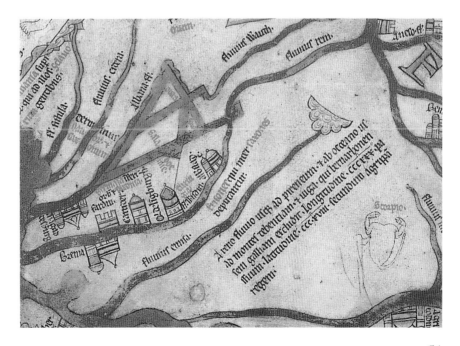

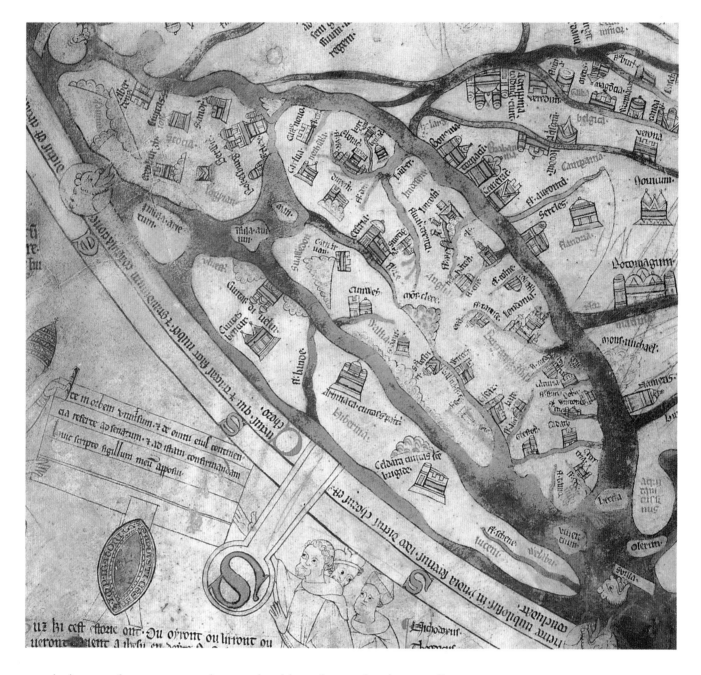

not in its actual sequence, and some should not be on the river at all. This is not the map's only quirk in Germany. Between the Weser and Elbe in the north and the Danube's tributaries in the south, is an inexplicable rectilinear pattern – there is nothing comparable anywhere else on the map. Clearly the map's draftsman mistook what he was copying, but what it was – or what he thought it was – is not clear. The lines are not drawn as mountains, though this is possibly what they were on the exemplar. They are not rivers, though the River Elbe hugs their eastern edge, for there is no trace of blue colouring. Miller suggests – but clearly himself without much conviction – that they may be intended as canals linking Elbe and Danube, or as the *limes Saxoniae*, the boundary area on Charlemagne's eastern frontier.

There is no sign that the map used itineraries in the British Isles, oddly, perhaps, since Matthew Paris certainly made use of them for his maps of Britain. The river system in England is closely related to what Matthew shows and must derive from the Roman world map, and, as on Matthew's maps, most English cathedral cities are shown – it was unfortunate that Hereford was originally one of four left out, along with Chichester, Norwich and Salisbury. As we have seen, the appearance of Carnarvon and Conway reflects current events when the map was being made, and in south-west England, Glastonbury and South Cadbury, supposed site of Camelot,[3] are a tribute to fashionable interest in the legends of King Arthur. Some places, such as Kirkham Abbey in Yorkshire, may have been shown because of some connection with Lincoln or with Richard of Holdingham himself that we know nothing of.[4] In Scotland, as on the River Weser in Germany, the relative positions of the towns, though not quite random, are far from reality. In fact, whatever use he may have made of itineraries for his map, we can be certain that neither Richard nor anyone else envisaged it as a guide to travel. Though a note on the Ebstorf map mentions that this was a possible use for a map, in practice no surviving medieval map earlier than the portolan charts was made to help anyone to get from one place to another.

This was one purpose the map was not meant to serve, but what its intended purpose was, what precisely it was meant to do either at Lincoln or at Hereford we can only guess or infer. What the map does today, however, is to show us how a cultured, well-read person in the England of Edward I might view the world – and, though the colours have faded and much of the visual impact is lost, it does this with a clarity and charm that we can still appreciate seven hundred years later.

THE BRITISH ISLES
Though badly distorted to fit within the circular frame, the coastal outline is in fact related to the much better outlines found on the Cotton world map and Matthew Paris's maps of Britain (pp.28, 36). The towns entered in Britain on the Hereford map could provide clues, not yet understood, to the connections or interests of the map's author. These might explain, for instance, why Kirkham Abbey in Yorkshire appears, or why St Davids is the only see entered in Wales.

NOTES

1 Both Miller, iv, p.33, and his facsimile of the map misread this inscription in trying to make sense of it; it is given correctly by Bevan and Phillott, p.74.
2 Bevan and Phillott, p.15; Miller, iv, p.7.
3 This is surely the identity of *Cadan*', neither a seriously misplaced Caen (as Bevan and Phillott, p.170; Miller, iv, p.18) nor a seriously miscopied Salisbury (as Crone 1954, p.9). Though its identification with Camelot is first recorded in the mid-sixteenth century (G. Ashe, *The Quest for Arthur's Britain* (London, 1968), p.156), it was probably a long-standing tradition.
4 That Kirkham was one of the many places that Edward I visited (Crone 1954, p.9) seems unlikely to be relevant.

APPENDIX 1
INSCRIPTIONS OUTSIDE THE FRAME OF THE MAP

Letters in italics in the transcriptions are expansions of words abbreviated on the map.

1 At the top in the Last Judgment scene

Scroll from the trumpet of the angel on the left
Leuez · si uendrez a ioie pardurable ·
> Rise – you will come to perpetual bliss.

Scrolls from the hands of Christ
Ecce testimonium meum ·
> Behold my witness.

Prayer of the Virgin Mary
Veici beu fiz mon piz · de deinz la quele chare preistes ·
E les mamelectes · Dont leit de uirgin queistes ·
Eyez merci de touz · si com uos memes deistes ·
Ke moi ont serui · Kant sauueresse me feistes ·
> See, dear son, my bosom within which you became flesh, and the breasts from which you sought the Virgin's milk. Have mercy, as you yourself have promised, on all those who have served me, who have made me their way to salvation.

Scroll from the trumpet of the angel on the right
Leuez · si alez au fu de enfer estable ·
> Rise – you are going to the fire prepared in hell.

2 In the bottom left corner

At the top
Lucas in euuangelio Exiit edictu*m* ab augusto cesare · ut describeretur huniuersus orbis ·
> In the Gospel of Luke: A decree went out from Caesar Augustus that the whole world should be described.

On the emperor's document
Ite in orbem vniue*r*sum · *et* de omni eius continencia referte ad senatum · *et* ad istam confirmandam Huic scripto sigillum meu*m* apposui ·
> Go into the whole world and report back to Senate on every continent – and to confirm this I have attached my seal to this document.

On the emperor's seal
S*IGILLVM* AUGVSTI · CESARIS · IMPIRATORIS ·
> Seal of Augustus Caesar, emperor.

Beside the surveyors
Nichodoxus · Theodocus · Policlitus ·
> Nichodoxus; Theodocus; Policlitus.

Below the emperor's seal
Tuz ki cest estorie ont · Ou oyront ou lirront ou ueront · Prient a ihesu en deyte · De Richard de haldingham o de Lafford eyt pite · Ki lat fet e compasse · Ki ioie en cel li seit done ·
> All those who possess this work – or who hear, read or see it – pray to Jesus in his godhead to have pity on Richard of Holdingham or of Sleaford, who made it and set it out, that he may be granted bliss in heaven.

3 In the bottom right corner

At the top
Descripcio orosii de ornesta mundi · sicut interius ostenditur
> Orosius's account of the *Ornesta* of the world, as shown within.

Above the huntsman
passe auant ·
> Go ahead.

4 On tabs spaced around the map

M O R S
> Death

5 Around the edge of the parchment

In the top left quarter
+ A ⦂ IULIO ⦂ CESARE ⦂ ORBIS ⦂ TERRARVM ⦂ METIRI ⦂ CEPIT ⦂
> Measuring the lands of the earth was undertaken by Julius Caesar.

In the top right quarter
+ A NICODOXO ⦂ OMNIS ⦂ ORIENS ⦂ DIMENSVS ⦂ EST ⦂
> All the east was measured by Nichodoxus.

In the bottom left quarter
+ A TEODOCO ⦂ SEPTEMTRION ⦂ ET ⦂ OCCIDENS ⦂ DIMENS*US* ⦂ EST ⦂
> The north and west were measured by Theodocus.

In the bottom right quarter
+ A ⦂ POLICLITO ⦂ MERIDIANA ⦂ PARS ⦂ DIMENSVS ⦂ EST ⦂
> The southern part was measured by Policlitus.

APPENDIX 2
REPRODUCTIONS OF THE MAP

Much of the scholarly work done on the Hereford map has taken as its basis one or other of the copies of the map, published or unpublished. The following are the principal reproductions of the map, with identifying features of those that are printed from drawn copies.

1 (a) Royal Geographical Society, London, Maps, World 448. Manuscript, full size, coloured. Copied 'by Thomas Ballard of Ledbury from the Original in the Chapter House Hereford Cathedral', 1831.

(b) Bibliothèque Nationale, Paris, Ge A.649. Manuscript, full size, coloured. Copied from 1(a), 1841.

(c) E.-F. Jomard, *Les monuments de la géographie ou recueil d'anciennes cartes européennes et orientales* (Paris, – Duprat et al., [1842-62]), no.XIV (as eventually issued in the bound volume; nos 1-12 in the original serial publication). Printed, full size, in six sheets, uncoloured lithograph. Copied by E. Rembielinski from 1(b), lithographed by – Kaeppelin. Identifying features: omits the names of 'Carnaruon' and 'Cunweh' in Wales, 'ciuitas bencur' and 'fl. bande' in Ireland; in the inscription referring to Orosius in the bottom right corner 'interius' is spelled 'iterius'.

1(b) and 1(c) are faithful copies of 1(a) and 1(b), but 1(a) is far from being a faithful copy of the original. Its most obvious error may be deliberate: the circle of the map itself has been enlarged, so that it protrudes beyond the outer border in the middle of each side and wholly interrupts the border at the bottom. This has consequential changes in the spacing of the inscription around the edge. But besides this, unfamiliar words have been miscopied (e.g. 'essedones' appears as 'Eitedones') and many names are omitted altogether.

2 (a) Photographs, reduced, of the whole map in four sections, uncoloured. By T. Ladmore of Hereford, 1868.

(b) Specimen section of the map circulated with a prospectus for 2(c), covering part of north-east Asia (the area between the *monoculi* at the top, the pelican bottom left and the inscription about Parthia bottom right). Printed, full size, single sheet, five-colour lithograph. Copied by G.C. Haddon of Hereford but apparently with reference also to 2(a), lithographed by E. Gailliard & Co. of Bruges, 1869 or 1870.[1] Identifying features: *ciuitas* is spelled 'ciuicas' after Cristoas, 'ciustas' after Octoricirus.

(c) Published by E. Stanford, London. Printed, full size, six sheets, five-colour lithograph. Edited by F.T. Havergal of Hereford, copied by G.C. Haddon of Hereford, F. Rogers of London and W. Dutton of Hereford but apparently with reference also to 2(a), lithographed by

E. Gailliard & Co. of Bruges, 1872. The area covered by 2(b) has been redrawn. Identifying features: in the area covered by 2(b) *montes* is spelled 'monies' before 'Paropanitates' (the Paro Pamisus Mountains); in Ireland the mountain beside Kildare is shaded with slightly curved, not U-shaped, lines; in the inscription along the edge in the bottom right corner the cross-bar of the T in 'POLICLITO' extends as far to the right as the lower part of the letter.

In what is printed in black and red, 2(c) is an extremely accurate copy of the map and it is difficult to find variations from the original that serve easily to identify it. The other colours, however, are not only a poor guide to the colours on the original but depart from it in detail, especially in failing to distinguish correctly the colours used for shading in the pictures on the map itself and in the borders.

3 K. Miller, *Mappaemundi: Die ältesten Weltkarten* (6 vols, Stuttgart, J. Roth, 1895-8), vol.4, inset. Printed, three-sevenths size, single sheet, three-colour lithograph. Edited by K. Miller, lithographed by G.F. Krauss of Stuttgart, 1896. Identifying features: in Scotland and England 'Muneth' and 'bathe' are spelled with *th*, not *þ*; in the inscription referring to Orosius in the bottom right corner 'interius' is spelled 'iterius' but with a line above the *i* standing for the missing *n*.

Though he recognised the inaccuracy of 1(c), Miller seems to have underestimated the accuracy of 2(c) while, unfortunately, accepting its colouring as correct. He describes how he used 1(c) for the outline of 3 – which thus has the same enlarged circle of the map – and arrived at the text of the inscriptions from a comparison of 1(c), 2(c) and Bevan and Phillott, entering the readings thus established on 3 as well as in the text of his book.[2] He lists inscriptions which he judged incorrect on the original map and which he has corrected both on 3 and in the text of his book; but he has made further alterations silently on both.[3] The letter-forms in the inscription around the edge of the map differ significantly from those of the original.

4 Royal Geographical Society, London, negative A19922. Photograph, reduced, of the whole map, uncoloured. By C.S. Priestley of Camberwell, c.1948.
Taken while the map was in London for conservation work.

5 *Reproductions of early maps III: The world map by Richard of Haldingham in Hereford Cathedral circa A.D.1285* (London, Royal Geographical Society, 1954). Printed, full size, nine sheets, collotype from photographs.
The first large-size photographic reproduction of the map,

and thus the first to be fully reliable. The detail, however, is not as sharply clear as might be wished.

6 (a) British Library, Manuscript Collections, Deposit 8754, Ektachromes 93528-93533. Photographs, 5×4 in. (127×102 mm.) transparencies, of the whole map (93528), of the whole map in four sections and of one detail, coloured. By British Library Photographic Service, 1989-90.

(b) British Library, Manuscript Collections, Deposit 8754, Kodachromes 58575-58588. Photographs, 35 mm. transparencies, of the whole map (58575) and of thirteen details, coloured. By British Library Photographic Service, 1989-90.

(c) British Library, Manuscript Collections, Deposit 8754, negatives 93537-93572, 93651. Photographs, 7×5 in. (178×127 mm.), of the whole map (93651), of the whole map in four sections and of thirty-two details, un-coloured. By British Library Photographic Service, 1989-90.

(d) British Library, Manuscript Collections, Deposit 8754, negatives 93573-93601. Photographs, ultra-violet light, 7×5 in. (178×127 mm.), of the whole map in twenty-nine overlapping sections, uncoloured. By British Library Photographic Service, 1989-90.

Although there is variation between different parts of the map and some details do not show up under the ultra-violet light, 6(d) is sharp and clear. Effectively, however, no more is visible under ultra-violet than under normal light.

7 Dean and Chapter of Hereford. Photograph, transparency, of the whole map, coloured. By G. Taylor, 1990.

Taken on the map's return to Hereford from the British Library.

8 (a) Hereford Mappa Mundi Trustee Co. Ltd. Photographs, 5×4 in. (127×102 mm.) transparencies, of the whole map and of the whole map in five sections, coloured. By M. Slingsby, 1995.

(b) Hereford Mappa Mundi Trustee Co. Ltd. Photographs, 60 mm. transparencies, of thirty details, coloured. By M. Slingsby, 1995.

These are the photographs used for the present book.

From some of these reproductions derive many others, often reduced in size. Thus Thomas Ladmore, photographer at Hereford, was offering for sale in 1873 photographs of 2(c) in three sizes: 13 by 11 inches, 11 by 9 inches and 5½ by 4½ inches.[4] Illustrations of the map in a single publication do not necessarily derive from the same original reproduction and it is often not made clear that illustrations are from a redrawn version and not from the original map. Thus the illustrations in Bevan and Phillott derive from 2(c) (frontispiece) and 2(a) (following p.xlvii); those in Moir and Letts from 2(b) (Asia showing camel, pelican, etc.) and 2(c) (all others); those in Harvey from 2(c) (pp.29, 33) and 6(a) (p.31).

The relationship between these reproductions and some derivatives is complicated. Thus, the reduced four-colour facsimile published in 1975 by A.E. Smith (Printers) Ltd of Gloucester takes as its basis a black-and-white photograph of 2(c), but prints in red the Red Sea and Persian Gulf, bands of decoration in the Last Judgment scene, the name of the Mediterranean and the inscription around the edge of the parchment, all of which have not been taken from 2(c) nor from a photograph of the original map but have been redrawn.

NOTES

1 *Transactions of the Woolhope Naturalists' Field Club* (1868), p.238; (1869), p.152.
2 Miller, iii, pp.4-5.
3 Miller, iii, p.5n. Among further alterations that Miller made silently on 3 and in his text are 'uiuunt' for 'inueni' in the inscription on the beasts like minotaurs, and the substitution of *th* for þ in 'Muneþ' and 'baþe'.
4 Bevan and Phillott, preface.

SELECT BIBLIOGRAPHY

Books giving a full description and discussion of the map:

BEVAN, W.L., and PHILLOTT, H.W., *Mediaeval Geography: An Essay in Illustration of the Hereford Mappa Mundi* (London, E. Stanford; Hereford, E.K. Jakeman; 1873; reprinted Amsterdam, Meridian, 1969)

MILLER, K., *Mappaemundi: Die ältesten Weltkarten* (Stuttgart, J. Roth; 6 vols, 1895-8), vol.4

CRONE, G.R., *Memoir* accompanying *Reproductions of early maps III: The world map by Richard of Haldingham in Hereford Cathedral circa A.D.1285* (London, Royal Geographical Society; 1954)

DESTOMBES, M., ed., *Mappemondes A.D.1200-1500* (Monumenta Cartographica Vetustioris Aevi, vol.1; *Imago Mundi*, supplement 4; Amsterdam, N. Israel; 1964), pp.197-202

MORGAN, N., *Early Gothic Manuscripts II: 1250-1285* (Survey of Manuscripts Illuminated in the British Isles, vol.4; London, Harvey Miller; 1988), pp.195-200

Brief guides to the map:

MOIR, A.L., and LETTS, M., *The World Map in Hereford Cathedral; The Pictures in the Hereford Mappa Mundi* (Hereford, Hereford Cathedral; [1955]; 7th edition 1975)

JANCEY, M., *Mappa Mundi: The Map of the World in Hereford Cathedral* (Hereford, Friends of Hereford Cathedral; 1987; new editions 1994, 1995)

Discussions of the map's author:

DENHOLM-YOUNG, N., 'The *mappa mundi* of Richard of Haldingham at Hereford', *Speculum*, 32 (1957), pp.307-14; reprinted in *Collected Papers of N. Denholm-Young* (Cardiff, University of Wales Press; 1969), pp.74-82

EMDEN, A.B., *A Biographical Register of the University of Oxford to A.D.1500* (Oxford, Clarendon Press; 3 vols, 1957-9), i, p.556 *sub* De la Bataylle, Richard

YATES, W.N., 'The authorship of the Hereford mappa mundi and the career of Richard de Bello', *Transactions of the Woolhope Naturalists' Field Club*, 41, part 2 (1974), pp.165-72

RAMSAY, N., 'Richard of Haldingham', in *The Dictionary of National Biography: Missing Persons*, ed. C.S. Nicholls (Oxford, Oxford University Press; 1993), pp.552-3

Details of the map's contents:

CRONE, G.R., 'New light on the Hereford map', *Geographical Journal*, 131 (1965), pp.447-62

GLENN, J., 'Notes on the *mappa mundi* in Hereford Cathedral', in *England in the Thirteenth Century: Proceedings of the 1984 Harlaxton Symposium*, ed. W.M. Ormrod (Woodbridge, Boydell and Brewer; 1986), pp.60-3

Discussions of the map's setting and significance:

BAILEY, M., 'The *mappa mundi* triptych: the full story of the Hereford Cathedral panels', *Apollo*, 137 (1993), pp.374-8

KUPFER, M., 'Medieval world maps: embedded images, interpretive frames', *Word and Image*, 10 (1994), pp.262-88

Books setting the map in the context of medieval cartography in general:

BEAZLEY, C.R., *The Dawn of Modern Geography* (London, J. Murray; 3 vols, 1897-1906; reprinted New York, P. Smith, 1949), especially ii, pp.465-642

HARLEY, J.B., and WOODWARD, D., eds, *The History of Cartography* (Chicago, University of Chicago Press; 6 vols, 1987-), i, pp.283-501

HARVEY, P.D.A., *Medieval Maps* (London, The British Library, 1991)

INDEX

Where the reference occurs only in the caption to an illustration the page number is in *italics*.